AS-Level

Biology

Edexcel

The Revision Guide

Editors:
Kate Manson, Rachel Selway, Julie Wakeling

Contributors:
Gloria Barnett, Claire Charlton, Martin Chester, Barbara Green, Anna-Fe Guy, Dominic Hall, Gemma Hallam, Becky May, Stephen Phillips, Claire Reed, Kate Redmond, Katherine Reed, Adrian Schmit, Emma Singleton, Sharon Watson.

Proofreaders:
Ben Aldiss, Vanessa Aris, James Foster, Tom Trust.

Published by Coordination Group Publications Ltd.

ISBN-10: 1 84146 956 4
ISBN-13: 978 1 84146 956 0

Groovy website: www.cgpbooks.co.uk
Jolly bits of clipart from CorelDRAW®
Printed by Elanders Hindson Ltd, Newcastle upon Tyne.

Contents

Section One — Biological Molecules

Section Two — Enzymes

Section Three — Cellular Organisation

Section Four — The Cell Cycle

Section Five — Exchanges With the Environment

Section Six — Transport Systems

Section Seven — Adaptations to the Environment

Section Eight — Sexual Reproduction

Section Nine — Ecosystems and the Environment

Water

Water — not the most stimulating topic to start the book with, you might say. But life can't exist without water. In fact boring, everyday water is one of the most important substances on the planet. Funny old world.

Water is Vital to Living Organisms

Water makes up about 80% of cell contents — it has loads of important **functions**, inside and outside cells.

1) Water is a **metabolic reactant**. That means it's needed for loads of important **chemical reactions**, like photosynthesis and hydrolysis reactions.
2) Water is a **solvent** — which means it can dissolve many substances. Most biological reactions take place **in solution**, so water's solvent properties are vital.
3) Water **transports** substances. The fact that it's a **liquid** and a solvent makes it easy for water to transport all sorts of materials around plants and animals, like glucose and oxygen.
4) Water helps with **temperature control**. When water **evaporates**, it uses up heat from the surface that it's on. This cools the surface and helps lower the temperature.

Water Molecules have a Simple Structure

The **structure of a water molecule** helps to explain many of its **properties**.
Examiners like asking you to relate structure to properties, so make sure you're clear on this.

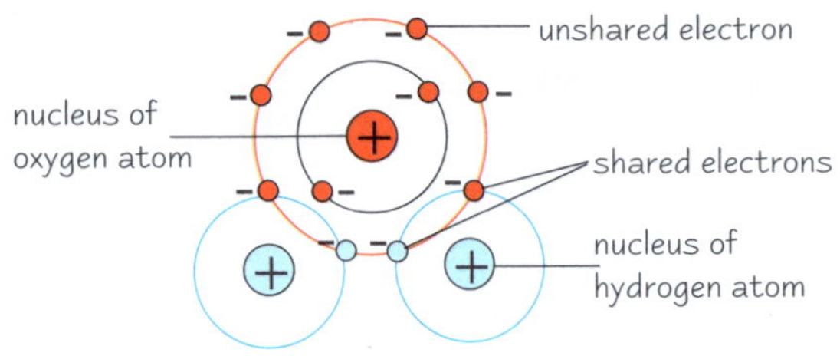

Water is **one atom of oxygen** joined to **two atoms of hydrogen** by **shared electrons**. Because the shared hydrogen electrons are pulled close to the oxygen atom, the other side of each hydrogen atom is left with a **slight positive charge**. The unshared electrons on the oxygen atom give it a **slight negative charge**. That means water is a **dipolar** molecule — it has negative charge on one side and positive charge on the other.

The **negatively charged oxygen atoms** of water **attract** the **positively charged hydrogen atoms** of other water molecules. This attraction is called **hydrogen bonding** and it gives water some of its special properties.

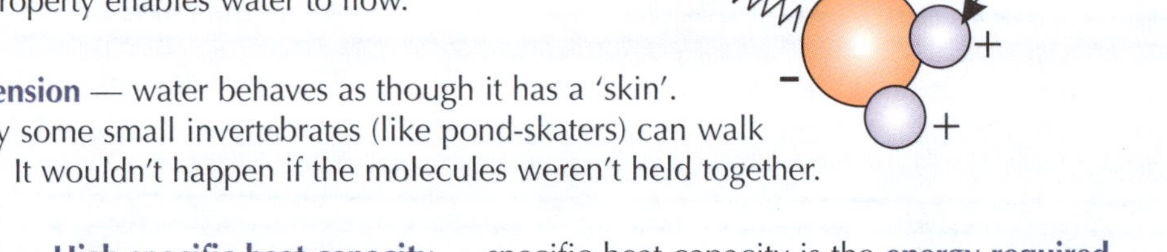

1) **Cohesion** — water molecules tend to stick together. This property enables water to flow.
2) **Surface tension** — water behaves as though it has a 'skin'. That's why some small invertebrates (like pond-skaters) can walk on water. It wouldn't happen if the molecules weren't held together.
3) **High specific heat capacity** — specific heat capacity is the **energy required** to raise the temperature of 1 gram of a compound by 1°C. Water has a high specific heat capacity — it takes a lot of energy to heat it up. This is useful for aquatic organisms, as it stops rapid temperature changes.
4) **High latent heat of vaporisation** — it takes a lot of heat to evaporate water, so it's great for cooling things.

Water

*Ice Floats — Which is **Useful***

Another weird property of water that's useful in nature is that it reaches its **maximum density** at about **4°C**. This means that ice is **less dense** than the water around it, so it **floats**. This acts as an **insulation** for the water below — so the sea or a lake won't freeze solid, which allows organisms under the ice to survive.

*Water's **Dipolarity** Makes it a **Good Solvent***

Because water is **dipolar**, it's a **good solvent** for other dipolar molecules. **Ionic** substances like salt, and **organic** molecules that have an **ionised group** will dissolve in water. The **positive** end of a water molecule will be attracted to a **negative ion** and the negative end of a water molecule will be attracted to a **positive ion**. The ions get **totally surrounded** by water molecules — in other words, they **dissolve**.

Remember — a molecule is dipolar if it has a negatively charged bit and a positively charged bit. This is also called "uneven charge distribution."

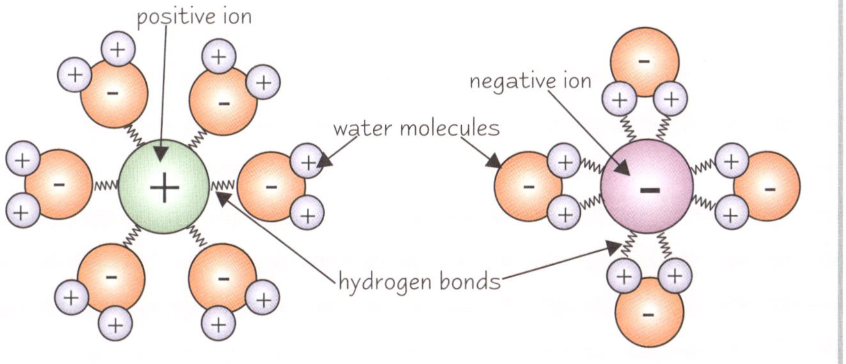

a) water molecules dissolving a positive ion

b) water molecules dissolving a negative ion

Chemical reactions take place much more easily in solutions than in solids, because the dissolved ions are **free to move around** and react.
Once dissolved, the solute can **easily be transported** by the water.

Practice Questions

Q1 State four uses of water in living organisms.

Q2 What is a 'dipolar molecule'?

Q3 Explain why the fact that ice floats on water is useful to living organisms.

Q4 State two reasons why water is useful as a solvent in living systems.

Exam Question

Q1 Relate the structure of the water molecule to its uses in living organisms. [12 marks]

Psssssssssssssssssssssssssssssssssssssss — need the loo yet?

Water is pretty darn useful really. It looks so, well, dull— but in fact it's scientifically amazing, and essential for all kinds of jobs — like maintaining aquatic temperatures, transporting things and enabling reactions. You need to learn all its properties and uses, plus the uses of inorganic ions. Right, I'm off — when you gotta go, you gotta go.

Carbohydrates

All carbohydrates contain only carbon, hydrogen and oxygen. Carbohydrates are dead important chemicals — they're the main energy supply in living organisms and some of them, like cellulose, have an important structural role.

Carbohydrates are Made from **Monosaccharides**

All carbohydrates are made from sugar molecules. A single sugar molecule is called a **monosaccharide**. Examples include glucose, ribose, deoxyribose, fructose (found in fruit) and galactose (present in lactose).

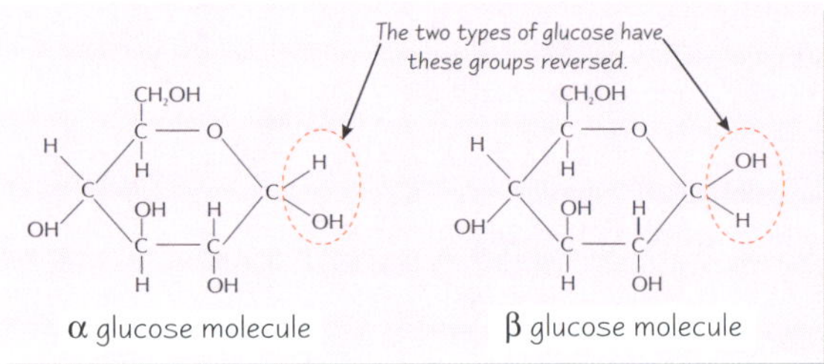

Remember, beta glucose has the H on the bottom as you look at the structural diagram.

ribose

deoxyribose

— pentose sugars

There are two types of glucose — **alpha** (α) and **beta** (β) glucose. You need to know how their molecules are arranged slightly differently. This has important effects on their **properties** and **functions** (see p.5).

Glucose is a **hexose sugar** — a monosaccharide with **six carbon** atoms in each molecule.

You also need to know the structures of **ribose** and **deoxyribose** (see p.10 for more), which are both **pentose** sugars. **Pentose sugars** are monosaccharides that have **five carbon atoms** in each molecule.

Disaccharides are **Two Monosaccharides** Joined Together

Disaccharides are sugars made from two monosaccharide sugar molecules stuck together. Examples are:

Disaccharide	Monosaccharides	Role
maltose	glucose + glucose	The sugar produced when starch is broken down
sucrose	glucose + fructose	The sugar transported around plants in the phloem
lactose	glucose + galactose	The main carbohydrate in milk

Fructose is the sweet sugar found in fruit. Galactose is added to some lipids and proteins to give certain glycolipids and glycoproteins needed in the body.

Glycosidic Bonds Join Sugars Together

Sugars are held together by **glycosidic bonds**. When the sugars join, a molecule of water is squeezed out. This is called a **condensation reaction**.

If you're asked to show a condensation reaction in an exam, don't forget to put the water molecule in as a product.

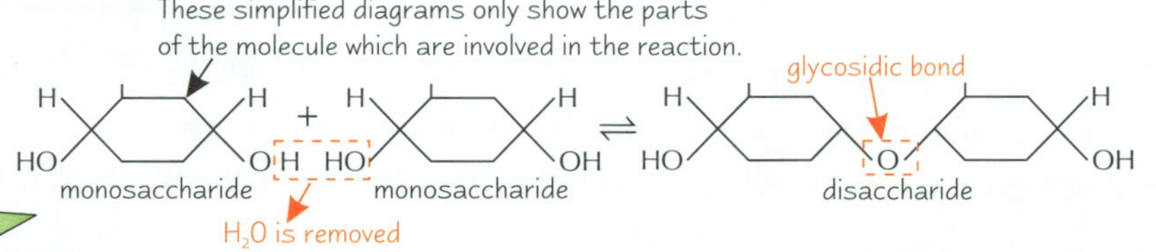

Hydrolysis Breaks Sugars Apart

When sugars are separated, the condensation reaction goes into **reverse**. This is called a **hydrolysis reaction** — a water molecule reacts with the glycosidic bond and breaks it apart.

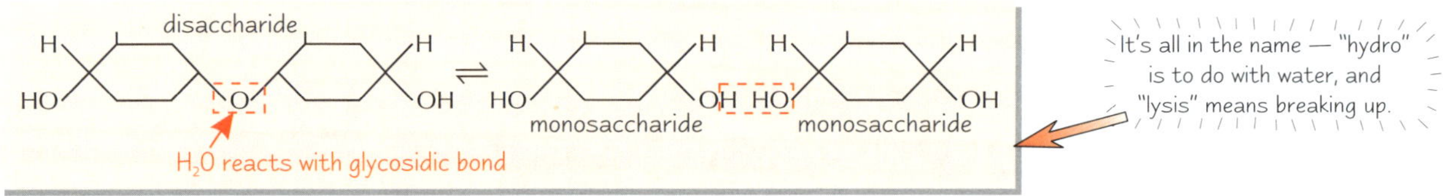

It's all in the name — "hydro" is to do with water, and "lysis" means breaking up.

Condensation and hydrolysis reactions are dead important in biology. **Proteins and lipids** are put together and broken up by them as well. So you definitely need to understand how they work.

Use the **Benedict's Test** for **Sugars**

The Benedict's test identifies **reducing sugars**. These are sugars that can donate electrons to other molecules — they include **all monosaccharides** and **some disaccharides**, e.g. maltose. When added to reducing sugars and heated, the **blue Benedict's reagent** gradually turns **brick red** due to the formation of a **red precipitate**.

To test for **non-reducing sugars** like sucrose, which is a disaccharide (two monosaccharides joined together), you first have to break them down chemically into monosaccharides. You do this by boiling the test solution with **dilute hydrochloric acid** and then neutralising it with sodium hydrogen carbonate before doing the Benedict's test.

Carbohydrates

Polysaccharides are Loads of Sugars Joined Together

Polysaccharides are molecules which are made up of **loads of sugar molecules** stuck together. The ones you need to know about are:

1) **starch** — the main storage material in plants;
2) **glycogen** — the main storage material in animals;
3) **cellulose** — the major component of cell walls in plants.

Examiners like to ask about the link between the structures of polysaccharides and their functions.

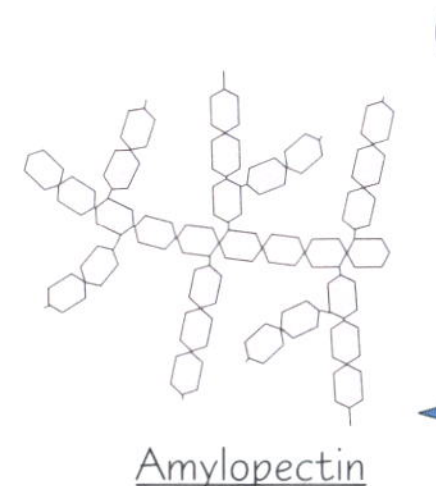

1. **Starch** is made up of **two** other polysaccharides of **alpha-glucose**:
   - **Amylose** is a long, **unbranched chain** of alpha-glucose. The angles of the glycosidic bonds give it a **coiled structure**, almost like a cylinder. Its **compact**, coiled structure makes it really **good for storage**.
   - **Amylopectin** is a long, **branched chain** of alpha-glucose. Its **side branches** make it particularly good for the storage of glucose — the enzymes that break down the molecule can get at the glycosidic bonds easily, to break them and release the glucose.

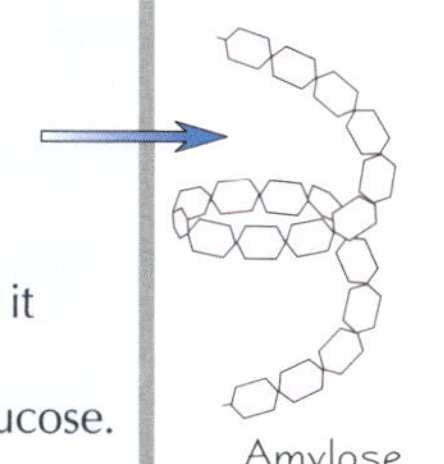

2. **Glycogen** is a polysaccharide of **alpha-glucose**. Its structure is very similar to amylopectin, except that it has **loads** more **side branches** coming off it. It's a very **compact** molecule found in animal liver and muscle cells. Loads of branches mean that stored glucose can be released quickly, which is **important for energy release** in animals.

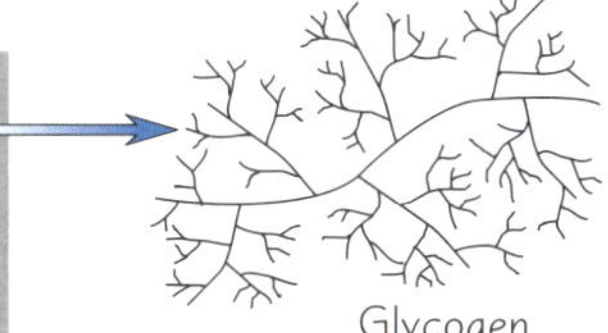

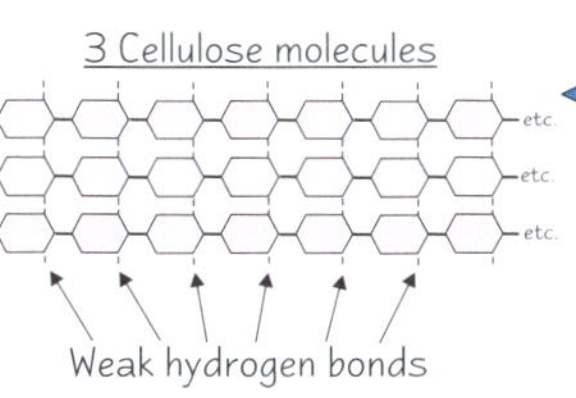

3. **Cellulose** is made of long, unbranched chains of **beta-glucose**. The bonds between beta sugars are **straight**, so the chains are straight. The chains are linked together by **hydrogen bonds** (see p.2) to form strong fibres called **microfibrils**. The strong fibres mean cellulose can provide **structural support** for cells. Another feature is that the **enzymes** that break the glycosidic bonds in starch can't reach the glycosidic bonds in cellulose, so those enzymes **can't break down cellulose**.

Use the Iodine Test for Starch

Make sure you always talk about iodine in potassium iodide solution, not just iodine.

In this test, you don't have to make a **solution** from the substance you want to test — you can use **solids** too. Dead easy — just add **iodine dissolved in potassium iodide solution** to the test sample. If there's starch present, the sample changes from **browny-orange** to a dark, **blue-black** colour.

Practice Questions

Q1 State the difference between a pentose and a hexose sugar.

Q2 What is the name given to the type of bond that holds sugar molecules together?

Q3 Explain the term "condensation reaction".

Q4 Name the two different types of molecule that are combined together in a starch molecule.

Q5 Name three polysaccharides and give the function of each one.

Q6 Cellulose is made from beta glucose. How does this help with its function as a structural polysaccharide?

Exam Questions

Q1 Describe how glycosidic bonds in carbohydrates are formed and broken in living organisms. [7 marks]

Q2 Compare and contrast the structures of glycogen and cellulose, showing how each molecule's structure is linked to its function. [10 marks]

Who's a pretty polysaccharide, then...

If you learn these basics it makes it easier to learn some of the more complicated stuff later on — 'cos carbohydrates crop up all over the place in biology. Remember that condensation and hydrolysis reactions are the reverse of each other — and don't forget that starch is composed of two polysaccharides. So many reminders, so little space ...

Lipids

Lipids are fats, oils and waxes — they are all made up of carbon, hydrogen and oxygen, and they're all insoluble in water. Ever seen a candle dissolve in water? No — exactly.

Lipids are Fats, Oils and Waxes — they're Useful

1) Lipids contain a lot of **energy per gram**, so they make useful **medium** or **long-term energy stores**. But they can't be broken down very quickly, so organisms use carbohydrates for **short-term storage**.
2) Lipids stored under the skin in **mammals** act as **insulation**. Skin loses heat from blood vessels, but the fatty tissue under the skin doesn't have an extensive blood supply, so it conserves heat.

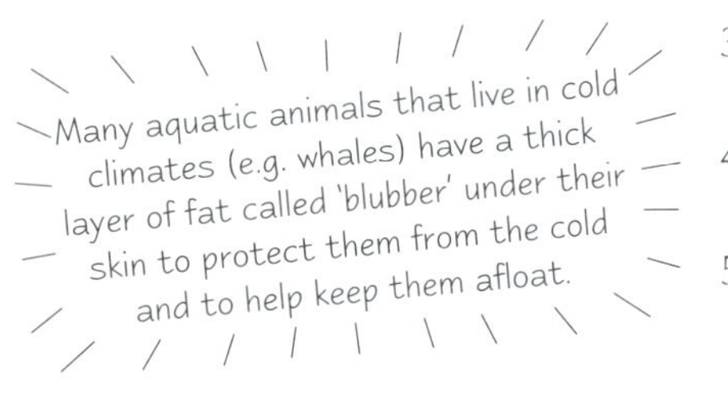

3) In **marine mammals** (e.g. whales, seals) lipids provide **buoyancy**, because the density of lipids is lower than that of muscle and bone.
4) Lipids under the skin and around the internal organs also provide **physical protection**, acting as a **cushion** against blows.
5) Lipids can act as a **waterproofing** layer — for example in the **waxy cuticle** on the surface of leaves and in the **exoskeleton** of insects. Lipids don't mix well with water, so water can't get through a lipid layer very easily.

Most Fats and Oils are Triglycerides

Most lipids are composed of compounds called triglycerides. Triglycerides are composed of one molecule of **glycerol** with **three fatty acids** attached to it.

Fatty acid molecules have long 'tails' made of **hydrocarbons**. The tails are '**hydrophobic**' (they repel water molecules). These tails make lipids insoluble in water. When put in water, fat and oil molecules **clump together** in globules to reduce the surface area in contact with water.

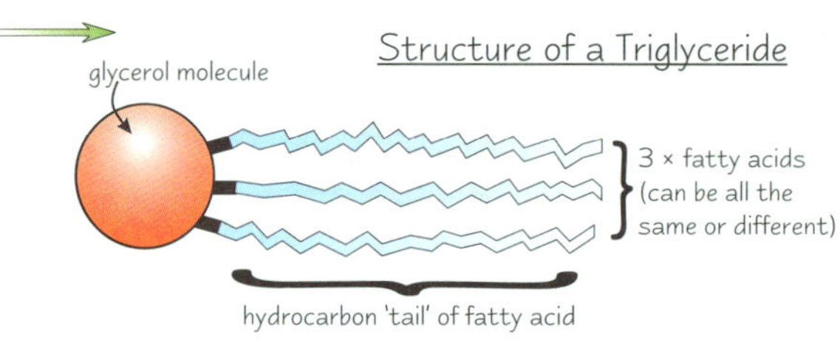

All **fatty acids** consist of the same basic structure, but the **hydrocarbon tail varies**. The tail is shown in the diagram with the letter 'R'.

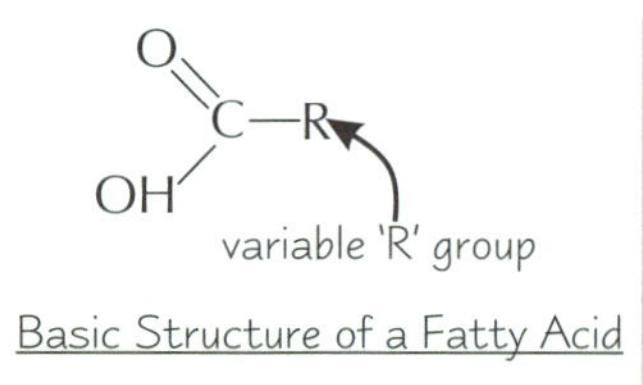

Triglycerides are Formed by Condensation Reactions

Like carbohydrates, lipids are formed by **condensation reactions** and broken up by **hydrolysis reactions**.

The diagram below shows a **fatty acid** joining to a **glycerol molecule**, forming an **ester bond**. A molecule of water is also formed — it's a **condensation reaction**. This process happens twice more, to form a **triglyceride**. The **reverse** happens in **hydrolysis** — a molecule of water is added to each ester bond to break it apart, and the triglyceride splits up into three fatty acids and one glycerol molecule.

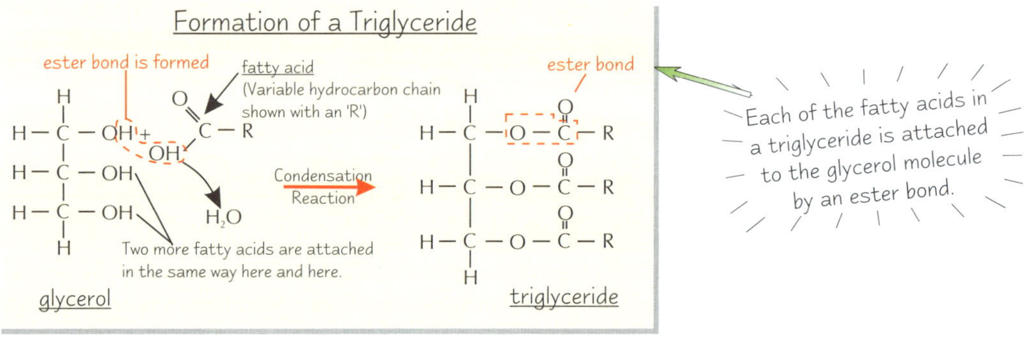

Each of the fatty acids in a triglyceride is attached to the glycerol molecule by an ester bond.

Lipids

*Lipids can be **Saturated** or **Unsaturated***

There are two kinds of lipid — **saturated** lipids and **unsaturated** lipids. **Saturated** lipids are mainly **animal fats** and **unsaturated** lipids are found mostly in **plants** (unsaturated lipids are called **oils**). The difference between these two types of lipids is in the **hydrocarbon tails** of their **fatty acids**.

1) Saturated fatty acids **don't** have any **double bonds** between their carbon atoms — every bond has a **hydrogen** atom attached. The lipid is 'saturated' with hydrogen.
2) Unsaturated fatty acids **do** have double bonds between carbon atoms. If they have **two or more** of them, the fat is called **polyunsaturated** fat.

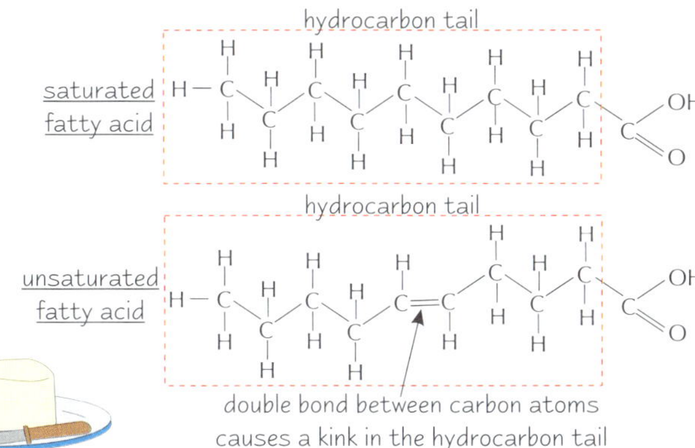

Unsaturated fats melt at lower temperatures than saturated ones. When used in margarine or butter spreads, it makes them easier to use 'straight from the fridge'.

***Phospholipids** are a Special Type of Lipid*

The lipids found in **cell membranes** aren't triglycerides — they're **phospholipids**. The difference is small but important:

1) In phospholipids, a **phosphate group** replaces one of the fatty acid molecules.
2) The phosphate group is **ionised**, which makes it **attract water** molecules.
3) So part of the phospholipid molecule is **hydrophilic** (attracts water) while the rest (the fatty acid tails) is **hydrophobic** (repels water).
4) Phospholipids form part of the **structure of cell membranes**. Their hydrophilic/hydrophobic properties make them good at this role (see p.26).

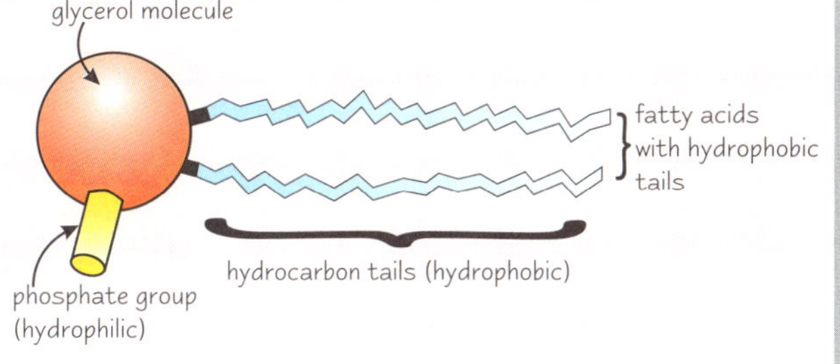

Practice Questions

Q1 Why is it wrong to call lipids 'fats'?

Q2 Why are lipids insoluble in water?

Q3 What's the name given to the type of bond that joins fatty acids to glycerol in a lipid molecule?

Q4 Explain the difference between a triglyceride and a phospholipid.

Exam Questions

Q1 State five functions of lipids in animals. For each, explain the feature of lipids that allows them to perform this function. [10 marks]

Q2 Describe the chemical reactions involved in the assembly and break down of triglycerides in living organisms. [8 marks]

Q3 Describe the differences between a triglyceride and a phospholipid, and explain how these differences affect the properties of the molecule. [8 marks]

What did the seal say to the upset whale? — Quit blubbering...

Truly awful joke — I hang my head in shame. You don't get far in life without extensive lard knowledge, so learn all the details on this page good and proper. Lipids pop up in other sections, so make sure you know the basics about how their structure gives them some quite groovy properties. Right, all this lipids talk is making me hungry — chips time...

Proteins

There are hundreds of different proteins — all of them contain carbon, hydrogen, oxygen and nitrogen. They are the most abundant organic molecules in cells, making up 50% or more of a cell's dry mass — now that's just plain greedy.

Proteins are Made from Long Chains of Amino Acids

All proteins are made up of amino acids joined together. All amino acids have a **carboxyl group** (-COOH) and an **amino group** (-NH_2) attached to a carbon atom.

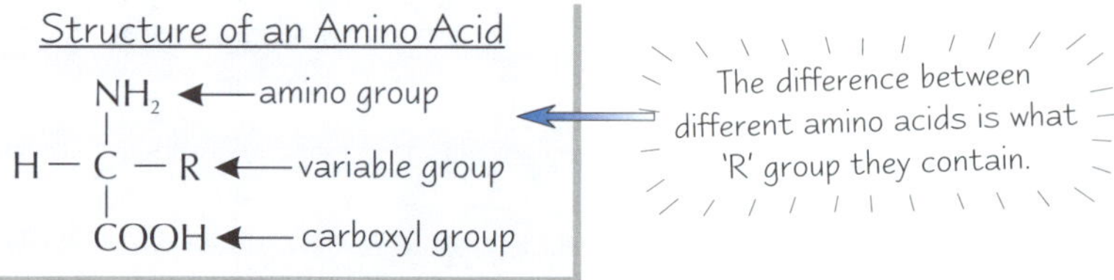

Proteins are Formed by...you guessed it...Condensation Reactions

Just like carbohydrates and lipids, the parts of a protein are put together by **condensation** reactions and broken apart by **hydrolysis** reactions. The bonds that are formed between amino acids are called **peptide bonds**. Two amino acids joined together are called a dipeptide. Lots of amino acids joined together are called a **polypeptide**.

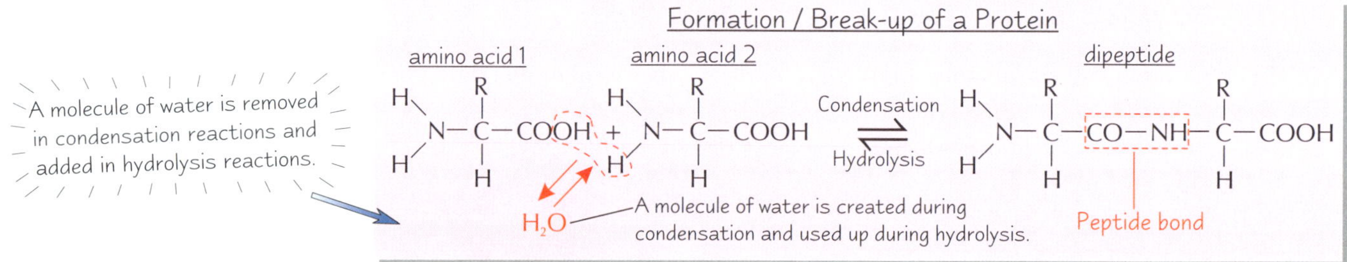

Proteins have up to Four Structures

Proteins are **big, complicated** molecules. They're easier to explain if you describe their structure in four 'levels'. These levels are called the protein's **primary, secondary, tertiary** and **quaternary** structures.

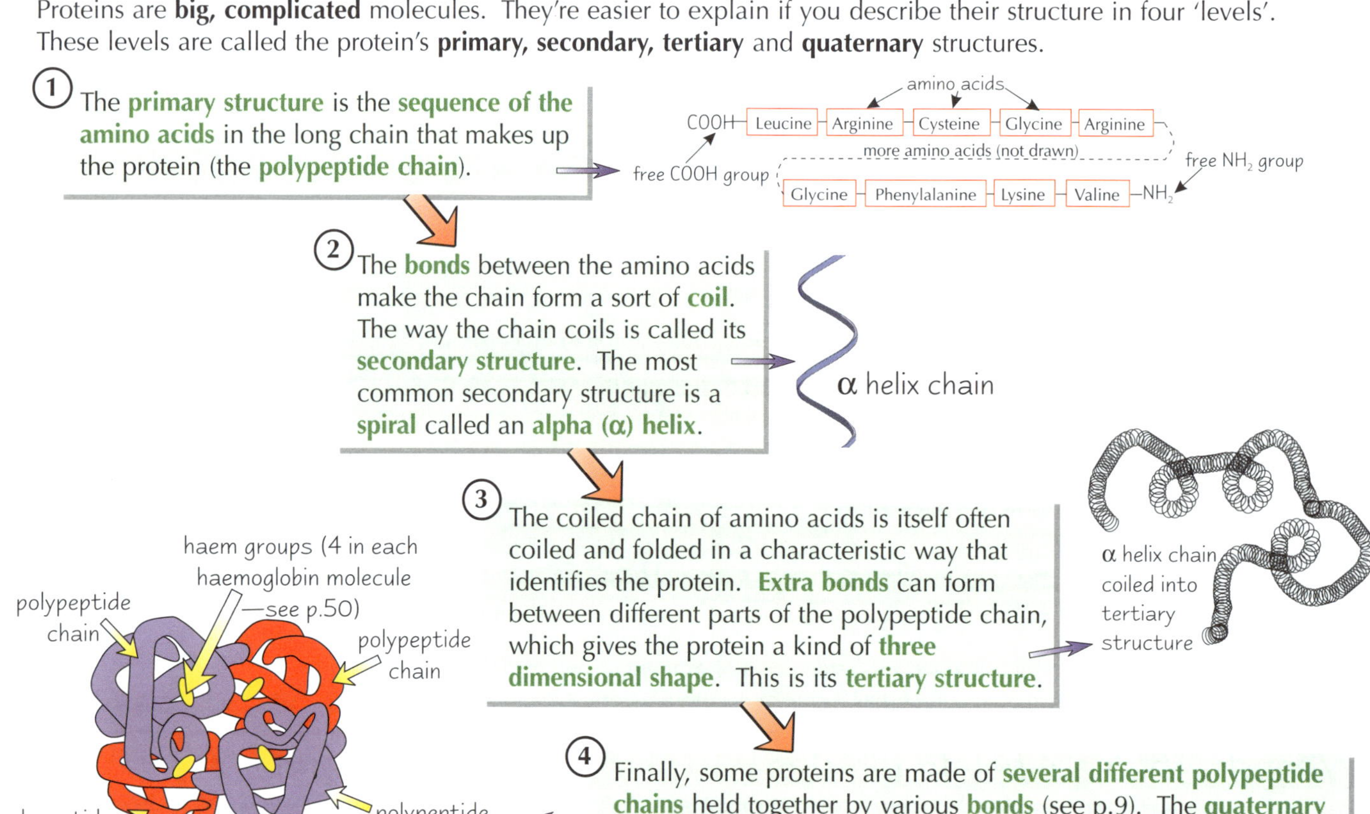

Proteins

Different Bonds Hold Proteins Together

Hydrogen bonds are weak bonds formed when a positively charged hydrogen atom is attracted by a negative charge on another atom.

Various types of bond hold protein molecules in shape.

1) **Hydrogen bonds** hold together the **secondary** structure of a protein. For example, in an alpha helix, a hydrogen bond forms between the **C=O** group of one amino acid and the **N-H** group of **another amino acid**, four amino acids along the polypeptide chain. Hydrogen bonds also help hold the tertiary structure of a protein together.
2) The **tertiary** structure of a protein is also held together by **weak ionic bonds**. These are weak attractions between a negatively charged part of one molecule and a positively charged part of another.
3) Whenever two molecules of the amino acid cysteine come close together, the sulphur in one cysteine bonds to the sulphur in the other cysteine. This is called a **disulphide bond**. It's part of the tertiary structure of a protein.

Protein Shape Relates to its Function

You need to learn a few **examples** of how proteins are **adapted for their jobs**.

Collagen is a 'fibrous protein' that forms **supportive tissue** in animals, so it needs to be strong.

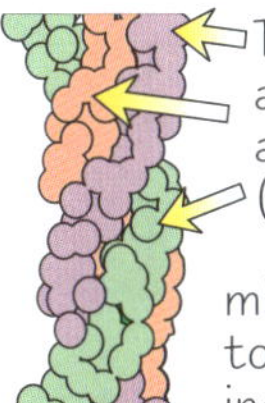

Insulin is a globular protein. It's a hormone that reduces blood glucose levels.

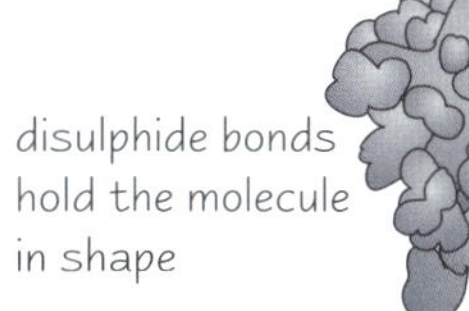

Enzymes are proteins — each enzyme has a specific shape, with an **active site** that locks onto the substrate molecule. See p.16 for more on enzymes and substrates.

Globular proteins are round and compact. They're soluble so they are easily transported around by the blood.

Use the Biuret Test for Proteins

There are **two stages** to this test.

1) The test solution needs to be **alkaline**, so first you add a few drops of **2M sodium hydroxide**.
2) Then you add some **0.5% copper (II) sulphate solution**. If a **purple layer** forms, there's protein in it. If it stays **blue**, there isn't. The colours are pale, so you need to look carefully.

Practice Questions

Q1 Name the four elements that are found in all proteins.

Q2 What are the common features of all amino acid molecules?

Q3 What is the name given to the bond that holds amino acids together in proteins?

Q4 Name three types of bond that hold a protein molecule in shape?

Exam Questions

Q1 Describe the structure of a protein, explaining the terms primary, secondary, tertiary and quaternary structure. No details are required of the chemical nature of the bonds. [9 marks]

Q2 Describe the structure of a collagen molecule, and explain how this structure relates to its function in the body. [6 marks]

The name's Bond — Peptide Bond...

Quite a lot to learn on these pages — proteins are annoyingly complicated.
Not happy with one, or even two, structures — they've got four of the things, and you need to learn 'em all.
Condensation and hydrolysis reactions are back by popular demand and you need to learn all the different bonds too.

Structure of DNA and The Genetic Code

*These pages are about the structure of DNA (**deoxy**ribonucleic acid) and RNA (plain ol' ribonucleic acid), plus a little thing called the genetic code, which is kinda important to us living things. (OK, spot the major understatement here.)*

DNA** and **RNA** are Very **Similar Molecules

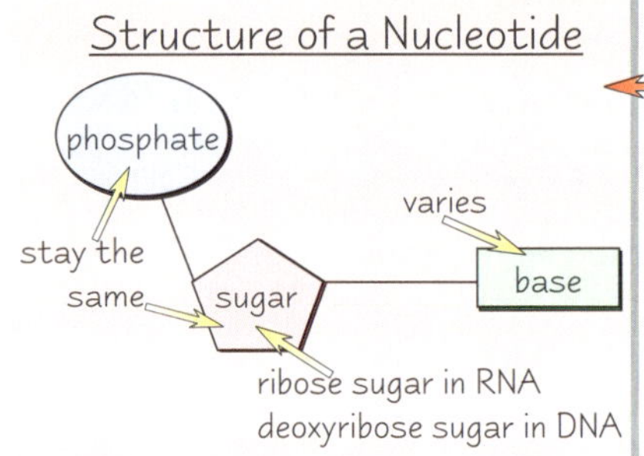

DNA and RNA are **nucleic acids** — made up of lots of **mononucleotides** joined together by condensation reactions. Mononucleotides are units made from a **pentose sugar** (with 5 carbon atoms), a **phosphate** group and a **base** (containing nitrogen and carbon) — these are also formed by condensation reactions.

The sugar in **DNA** mononucleotides is a **deoxyribose** sugar — in **RNA** mononucleotides it's a **ribose** sugar. Within DNA and RNA, the sugar and the phosphate are the same for all the mononucleotides. The only bit that's different between them is the **base**. There are five possible bases and they're split into two groups:

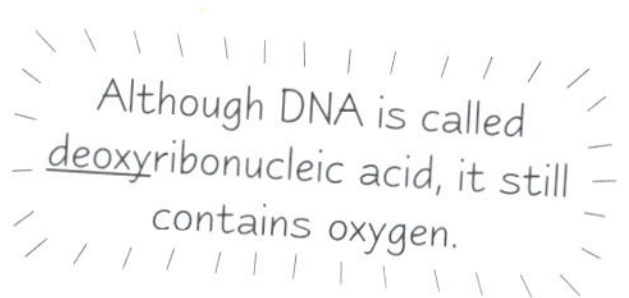

NAME	BASIC STRUCTURE	BASE	found in DNA	found in RNA
Purine bases	2 rings of atoms	**adenine**	✓	✓
		guanine	✓	✓
Pyrimidine bases	single ring of atoms	**cytosine**	✓	✓
		thymine	✓	✗
		uracil	✗	✓

DNA** and **RNA** are Polymers of **Mononucleotides

Mononucleotides join together by a **condensation reaction** between the **phosphate** of one group and the **sugar** molecule of another. As in all condensation reactions, **water** is a by-product.

DNA is made of **two polynucleotide strands**. **RNA** has just the one strand. In DNA, the strands spiral together to form a **double helix**. The strands are held together by **hydrogen bonds** between the bases.

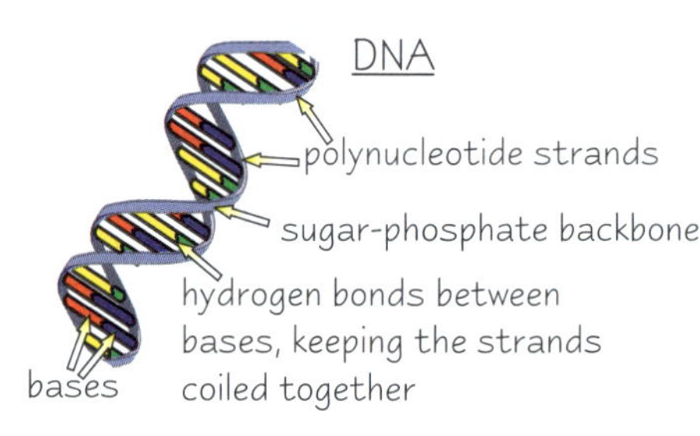

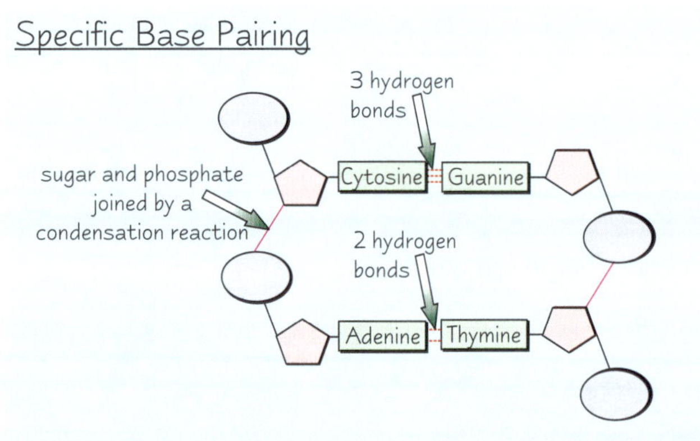

Each base can only join with one particular partner — this is called **specific base pairing**.

1) In DNA **adenine** always pairs with **thymine** (**A - T**) and **guanine** always pairs with **cytosine** (**G - C**).
2) It's the same in RNA, but **thymine**'s replaced by **uracil** (so it's **A - U** and **G - C**).

2 hydrogen bonds form between adenine and thymine.
3 hydrogen bonds form between guanine and cytosine.

DNA's Structure** Makes it **Good at its Job

1) The job of DNA is to carry **genetic information**. A DNA molecule is very, very **long** and is **coiled** up very tightly, so a lot of genetic information can fit into a **small space** in the cell nucleus.
2) Its **paired structure** means it can **copy itself** — this is called **self-replication** (see p.12). It's important for cell division and for passing on genetic information to the next generation.

Structure of DNA and The Genetic Code

DNA Contains the Basis of the Genetic Code

Genes are sections of DNA that code for a specific **sequence of amino acids** that forms a particular **protein**. The way that DNA codes for proteins is called the **genetic code**.

1) Genes code for specific amino acids with sequences of three bases, called **base triplets**. Different sequences of bases code for different amino acids. For example AGA codes for serine and CAG codes for valine.

2) There are **64** possible **base triplet combinations**. There are only about **20** amino acids in human proteins so there are some base triplets to spare. These aren't wasted though:
   - some amino acids use more than one base triplet.
   - some base triplets act as 'punctuation' to stop and start production of an amino acid sequence. These create **stop codons** and **start codons**.

The Genetic Code is Non-Overlapping and Degenerate

1) In the genetic code, each base triplet is read in sequence, separate from the triplet before it and after it. Base triplets **don't share** their **bases** — so the code is described as **non-overlapping**.

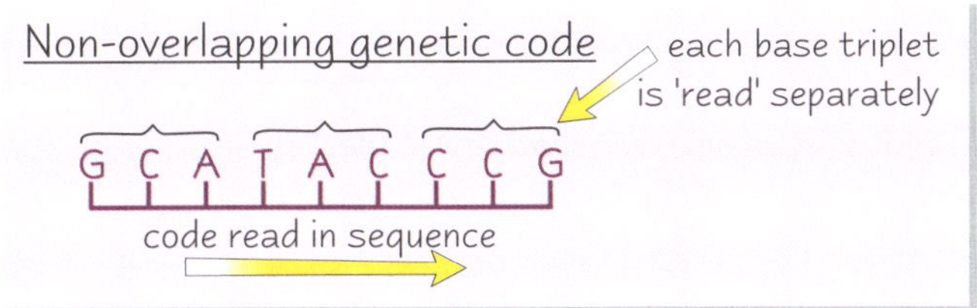

2) The genetic code in DNA is also described as **degenerate**. This is because there are **more triplet codes** than there are amino acids. Some **amino acids** are coded for by **more than one base triplet**, e.g. Tyrosine can be coded for by TAT or TAC.

3) There are sections of DNA that **don't code** for amino acids — these lengths of DNA are called **introns** (all the bits that do form part of the genetic code are called **exons**). Introns are removed from DNA during protein synthesis. Their purpose isn't known for sure.

Practice Questions

Q1 What three things are nucleotides made from?

Q2 Which base pairs join together in a DNA molecule?

Q3 What type of bonds join the bases together?

Q4 What is an intron?

Exam Questions

Q1 Explain how the structure of DNA is related to its function. [2 marks]

Q2 Describe, using diagrams where appropriate, how nucleotides join together and how two single strands of DNA become joined. [5 marks]

Give me a D, give me an N, give me an A! What do you get? — very confused…

*You need to know the basic structure of DNA and RNA and also how DNA's structure makes it good at its job. Then there's the genetic code to get to grips with — hmmm, rather you than me, but it **basically** comes down to the sequence of **bases**. I'm afraid there's nowt else you can do except buckle down, pull your socks up and get all them facts learnt.*

DNA Self-Replication and RNA

Here comes some truly essential stuff — DNA replication is the real nitty-gritty of biology. So eyes down for some serious fact-learning. I'm afraid it's all horribly complicated — all I can do is keep apologising. Sorry.

*DNA can Copy Itself — **Self-Replication***

DNA has to be able to **copy itself** before **cell division** can take place, which is essential for growth and development and reproduction — pretty important stuff.

1) **Specific base pairing** means that each type of base in DNA only pairs up with one other type of base — **A** with **T**, **C** with **G**.
2) When a molecule of **DNA splits**, the **unpaired bases** on each strand can match up with complementary bases on **free-floating mononucleotides** in the cytoplasm, making an **exact copy** of the DNA on the other strand. This happens with the help of enzymes.
3) The result is **two molecules** of DNA **identical** to the **original molecule** of DNA:

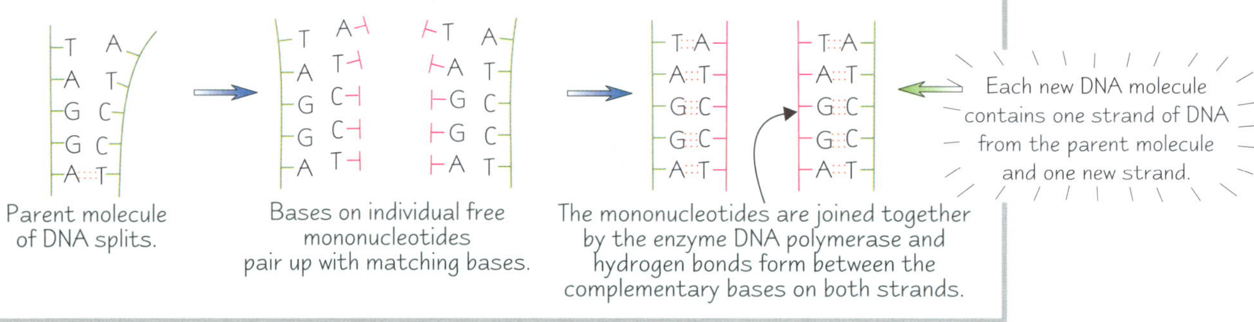

This type of copying is called **semi-conservative replication** — because **half** of the new strands of DNA are from the **original** piece of DNA.

There** are **Three Types** of **RNA

There are **three types** of RNA, and all are involved in **making proteins** (see p.14 for more on this — bet you can't wait).

Transfer RNA (tRNA)** has a **Binding Site** and an **Anticodon

1) **tRNA** is a **single polynucleotide strand** that's folded into a **clover shaped molecule**.
2) Each tRNA molecule has a **binding site** at one end, where a specific **amino acid** attaches itself to the bases there.
3) Each tRNA molecule also has a specific sequence of **three bases** at one end of it, called an **anticodon**.
4) The significance of binding sites and anticodons are all revealed over the page. But you need to know **where they are found** on a tRNA molecule, so learn the diagram on the right off by heart.

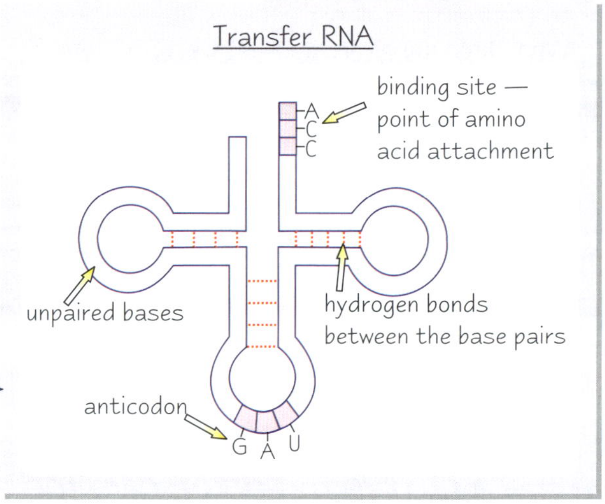

Transfer RNA

binding site — point of amino acid attachment

A C C

unpaired bases

hydrogen bonds between the base pairs

anticodon

G A U

DNA Self-Replication and RNA

Messenger RNA (mRNA) is a Reverse Copy of a DNA Strand

1) **mRNA** is a **single polynucleotide strand** that's formed in the **nucleus**.
2) The important thing to know about it is that it's formed by using a section of a **single strand of DNA** as a **template**. **Specific base pairing** means that mRNA ends up being an exact **reverse copy** of the DNA template section (see the piccy on the right to make sense of this).
3) You also need to know that the **3 bases in mRNA** that pair up with a base triplet on the DNA strand are called a **codon**. Codons are dead important for making proteins (see p.14), so **remember this word**. Make sure you realise that a codon has the **opposite bases** to a base triplet (except the base **T** is replaced by **U** in **RNA**).

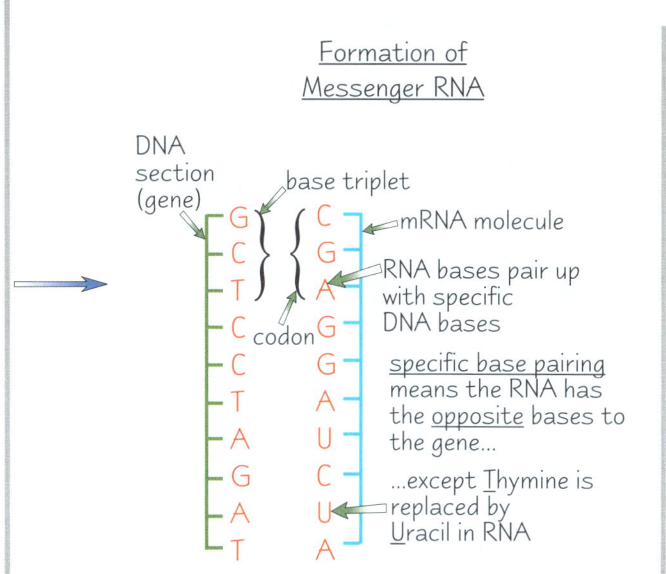

Ribosomal RNA (rRNA) — Ribosomes are the Site of Protein Synthesis

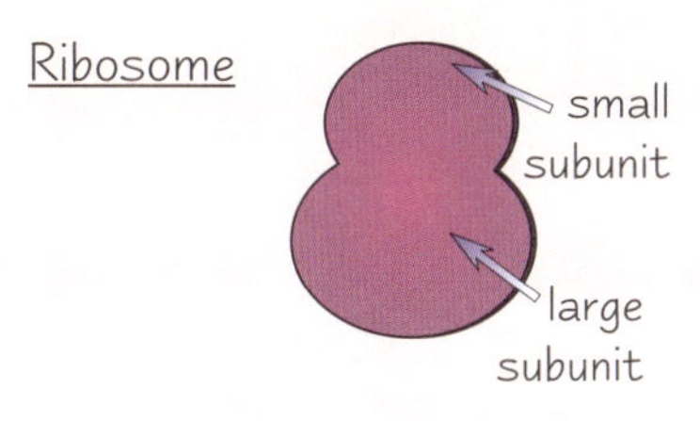

1) rRNA is made up of polynucleotide strands that are folded and attached to proteins to make things called **ribosomes**.
2) Ribosomes are the site where proteins are made — and that's what the next page is all about...

Practice Questions

Q1 What is the name used to describe the type of replication in DNA?

Q2 What three types of RNA are there?

Q3 What is the name of the group of three bases on mRNA that correspond to a base triplet on DNA?

Q4 What shape does a chain of tRNA fold itself into?

Exam Question

Q1 Describe and explain the semi-conservative method of DNA replication. [6 marks]

My genes are degenerate — there's a hole in the back pocket... *(I'll get my code)*

Quite a few terms to learn here — you're on the inescapable road to science geekville I'm afraid, and it's a road lined with crazy diagrams and strange words. DNA self-replication is sooo important — so make sure you understand what's going on. You need to learn the structure of the three types of RNA — it'll help you understand protein synthesis, on the next page.

Protein Synthesis

You've learnt about the genetic code and types of RNA — now you can learn all about its role in making proteins. This stuff is biology at its most clever, and it's probably going on inside you right now. Weird.

Protein synthesis (making proteins) happens in **two stages** — **transcription** and **translation**. It involves both DNA and RNA.

First Stage — *Transcription* Occurs in the *Nucleus*

Don't forget that, in RNA, adenine pairs up with uracil, not thymine.

In **transcription** a '**negative copy**' of a **gene** is made. This copy is called **mRNA**.

1) A gene (a section of DNA) in the DNA molecule **uncoils** and the hydrogen bonds between the two strands in that section break, separating the strands.
2) One of the strands is then used as the **template** for transcription — it's called the '**sense strand**'.
3) Free **RNA nucleotides** in the nucleus line up alongside the template strand. Once the RNA nucleotides have **paired up** with their **complementary bases** on the DNA strand they're joined together by the enzyme **RNA polymerase**.
4) The strand that's formed is **mRNA**.
5) It then moves out of the nucleus through a nuclear pore, and attaches to a **ribosome** in the cytoplasm, where the next stage of protein synthesis takes place.
6) When enough mRNA has been produced, the uncoiled strands of DNA re-form the hydrogen bonds and **coil back into a double helix**, unaltered.

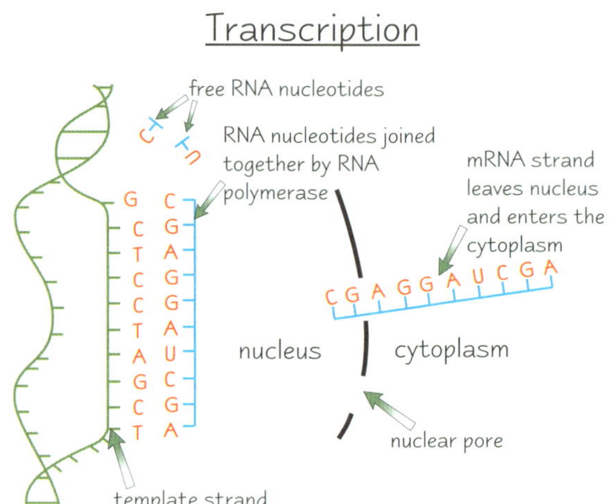

Second Stage — *Translation* Occurs in a *Ribosome*

In **translation**, **amino acids** are stuck together to make a **protein**, following the order of amino acids coded for on the mRNA strand.

1) The **mRNA strand** has travelled to a ribosome in the cytoplasm, and attached itself.
2) All 20 **amino acids** needed to make human proteins are in the cytoplasm. tRNA molecules attach to the amino acids and transport them to the ribosome.
3) In the ribosome, a tRNA molecule binds to the start of the mRNA strand. This tRNA molecule has the **complementary anticodon** to the **first codon** on the mRNA strand, and attaches by **base pairing**. Then a second tRNA molecule attaches itself to the **next codon** on the mRNA strand in the **same way**.
4) The two amino acids attached to the tRNA molecules are joined together with a **peptide bond** (using ATP and an enzyme).
5) The first tRNA molecule then **moves away** from the ribosome, leaving its amino acid behind. The mRNA then **moves across** the ribosome by one codon and a third tRNA molecule binds to the **next codon** that enters the ribosome.
6) This process continues until there's a **stop codon** on the mRNA strand that doesn't code for any amino acid. You're left with a line of amino acids joined by peptide bonds. This is a **polypeptide chain** — the **primary structure** of a protein. The polypeptide chain moves away from the ribosome and translation is complete.

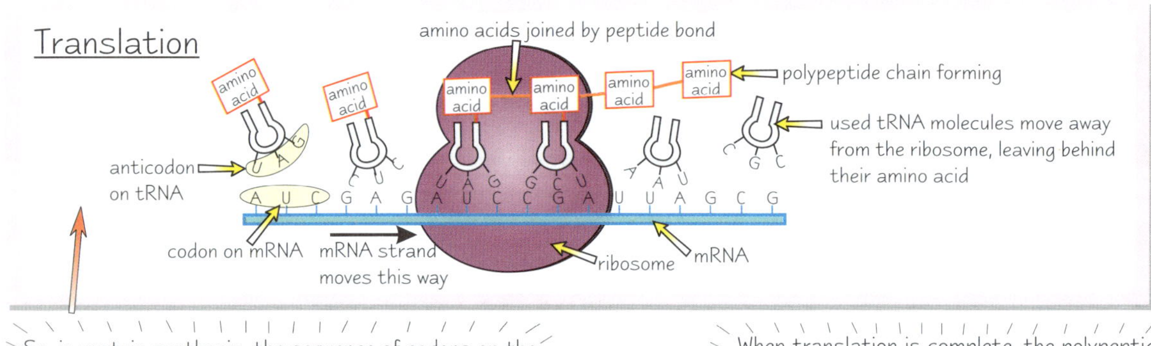

So, in protein synthesis, the sequence of codons on the mRNA strand determines the sequence of amino acids that makes up the primary structure of the protein.

When translation is complete, the polypeptide chain folds itself into its secondary and tertiary structure, and a protein is formed (see p.8).

Protein Synthesis

The Structure of an Enzyme is Determined by Protein Synthesis

1) All **enzymes** are **proteins**, which are sequences of amino acids.
2) The amino acid sequences are determined by the base sequence in DNA, so **DNA** determines the **structure of enzymes**.
3) Enzymes speed up all our **metabolic pathways** (see p.16). They have a big influence over how our **genes** are **expressed physically**, by controlling chemical reactions required for growth and development. This physical expression of the gene contributes to the organism's **phenotype** (what the organism looks like).

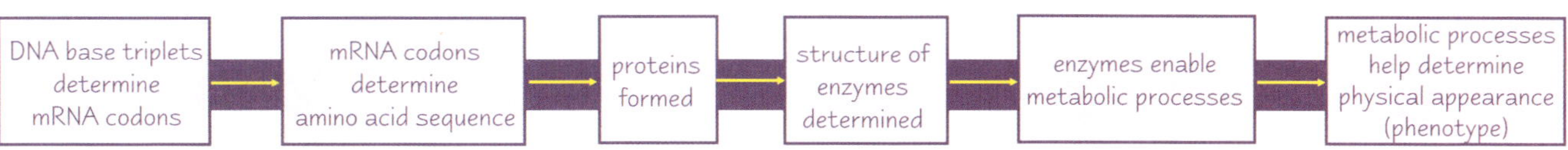

The Human Genome Project has Identified all Human Genes

The **Human Genome Project** is an international project that aimed to **map** where every human gene is found on our chromosomes. This goal was achieved — and is one of the most amazing scientific achievements ever. To extend the project, scientists are now trying to work out **what each gene codes for**. Knowing where a particular gene is found helps scientists learn to understand how they might be **damaged** or how **mutations** on some genes can lead to illness or disorders. It also makes it easier to replace faulty genes. You see, biology wasn't just invented to make your lives miserable — it's actually quite useful.

Here are some of the ways that the Human Genome Project might be of use in the future:

1) We might be able to change our **genetic inheritance** by eliminating genes that code for inherited diseases.
2) Doctors can make more **accurate diagnoses** of diseases.
3) Better drugs can be made to **target** specific diseases.
4) Drugs with **fewer side-effects** can be made.
5) There'll be more information to base **medical research** on.

Many people have ethical arguments against changing our natural genetic make-up. Some scientists are attempting to clone humans and it could be possible in the future to choose a 'designer' baby, with the genetic make-up of your choice. Many people dislike the idea of this level of interference.

Practice Questions

Q1 What are the two main stages in protein synthesis?

Q2 Which base is not present in RNA?

Q3 Write the RNA sequence which would be complementary to the following DNA sequence.
AATTGCGCCCG

Q4 Where does transcription take place?

Q5 Where does translation take place?

Exam Questions

Q1 Explain the terms codon and anticodon. [2 marks]

Q2 Describe the process of protein synthesis. [10 marks]

mRNA codons join to tRNA anticodons?! — I need a translation please…

When you first go through protein synthesis it might make approximately no sense, but I promise its bark is worse than its bite. All those strange words disguise what is really quite a straightforward process — and the diagrams are dead handy for getting to grips with it. Keep drawing them yourself, 'til you can reproduce them perfectly.

Action of Enzymes

*Enzymes crop up loads in biology — they're really useful 'cos they make reactions work more quickly. So, whether you feel the need for some speed or not, read on — because you **really** need to know this basic stuff about enzymes.*

Enzymes are Biological Catalysts

A catalyst is a substance that speeds up a chemical reaction without being used up in the reaction itself.

Enzymes speed up chemical reactions by acting as **biological catalysts**.

1) They catalyse every **metabolic reaction** in the bodies of living organisms. Even your **phenotype** (physical appearance) is down to enzymes that catalyse the reactions that cause growth and development.
2) Enzymes are **globular proteins** (see p.9) although some have **non-protein components** too.
3) Every enzyme has an area called its **active site**. This is the part that connects the enzyme to the substance it interacts with, which is called the **substrate**.
4) Enzymes are **specific**. For the enzyme to work, the substrate has to **fit** into the **active site**. If the substrate's shape doesn't match the active site's shape, then the reaction won't be catalysed. This means that enzymes work with very few substrates, usually only one.

Dave knew Sara was a lovely girl, but just couldn't get past the shape incompatibility thing

Enzymes Reduce Activation Energy

In a chemical reaction, a certain amount of energy needs to be supplied to the chemicals before the reaction will start. This is called the **activation energy** — it's often provided as **heat**.

Enzymes **reduce** the amount of activation energy needed, often making reactions happen at a **lower temperature** than they could without an enzyme. This **speeds** up the **rate of reaction**.

Graph Showing How Enzymes Speed up the Rate of Reaction

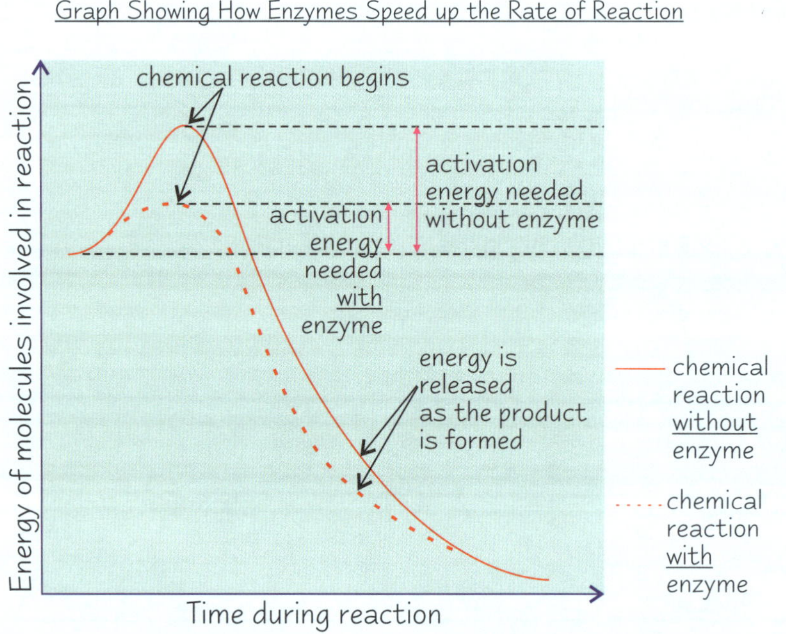

This happens because of the **enzyme-substrate complex**, which forms when a substrate fits into the active site:

1) If two substrate molecules need to be **joined**, attaching to the enzyme holds them **close together**, **reducing** any **repulsion** between the molecules so they can bond more easily.
2) If the enzyme is catalysing a **breakdown reaction**, fitting into the active site puts a **strain** on bonds in the substrate. This strain means the substrate molecule **breaks up** more easily.

Action of Enzymes

*Enzyme Activity can be **Inhibited***

Enzyme activity can be prevented by **enzyme inhibitors**. These molecules **bind to the enzyme** and interfere with the formation of the enzyme-substrate complex.

Inhibition can be **active site-directed** (competitive) or **non-active site-directed** (noncompetitive).

1) **Active site-directed inhibitors** have a **similar shape to the substrate**. They compete with the substrate to bond to the active site, but no reaction follows. Instead they **block** the active site, so **no substrate** can **fit** in it. How much inhibition happens depends on the **relative concentrations** of inhibitor and substrate — if there's a lot of the inhibitor, it'll take up all the active sites and stop any substrate from getting to the enzyme.

substrate (similar shape to inhibitor)

enzyme

inhibitor fits into active site

2) **Non-active site-directed inhibitors** bond to the enzyme **away from its active site**, but this causes the active site to **change shape**. They don't 'compete' with the substrate because even if there's a substrate in the active site, the inhibitor can still fit on.

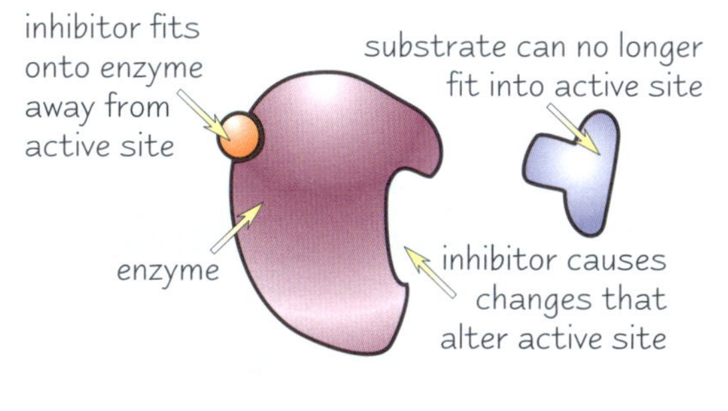

Competitive and noncompetitive inhibitors can be **reversible** or **non-reversible**.
This mainly depends on the **strength of the bond** between the enzyme and the inhibitor.

1) If it's a **strong**, **covalent bond** then the inhibitor can't be removed easily and the inhibition is **irreversible**.
2) If it's a **weaker hydrogen bond** or a weak, **intermolecular ionic bond**, then the inhibitor can be removed and the inhibition is **reversible**.

Practice Questions

Q1 Define the term "catalyst".

Q2 What is the name given to the amount of energy needed to start a reaction?

Q3 What is an "enzyme-substrate complex"?

Q4 What is the difference between a competitive and a noncompetitive enzyme inhibitor?

Exam Question

Q1 When a small amount of chemical X is added to a mixture of an enzyme and its substrate, the formation of reaction products is reduced. State, giving a reason, the likely nature of chemical X. [2 marks]

But why is the enzyme-substrate complex?

OK, nothing too tricky here. The main things to remember are that enzymes are a bit picky — each one will only work with the specific substrates that fit its shape — and that the energy needed for a reaction is less with an enzyme. Oh, and be able to describe the different types of inhibition. Go on, don't be shy ...

Factors that Affect Enzyme Activity

Just when you thought you'd seen the last of enzymes, here they are again to brighten up your day one more time. This time it's all about the conditions that enzymes work best in.

Temperature has a Big Influence on Enzyme Activity

Like any chemical reaction, the rate of an enzyme-controlled reaction increases when the temperature's raised. More heat means more **kinetic energy**, so molecules move faster. This makes the enzyme more likely to **collide** with the substrate. But, if the temperature increases beyond a certain point, the **reaction stops**. This is because the rise in temperature also makes the enzyme's particles **vibrate**:

1) If the temperature goes above a certain level, this vibration **breaks** some of the **bonds** that hold the enzyme in shape.
2) The **active site changes shape** and the enzyme and substrate **no longer fit together**.
3) At this point, the enzyme is **denatured** — it no longer functions as a catalyst.

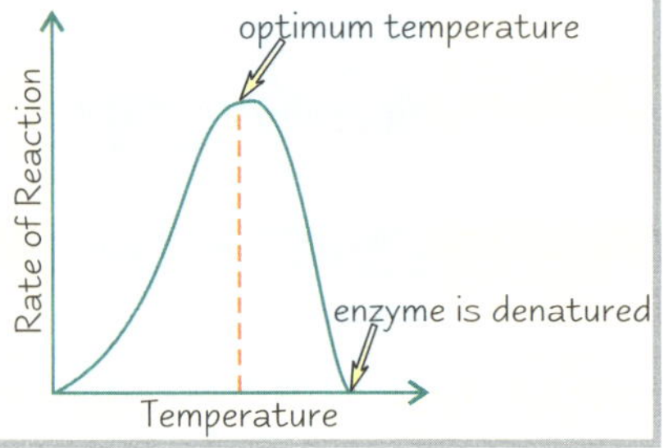

Every enzyme has an optimum temperature. In humans it's around 37°C but some enzymes, like those used in biological washing powders, can work well at 60°C.

pH Also Affects Enzyme Activity

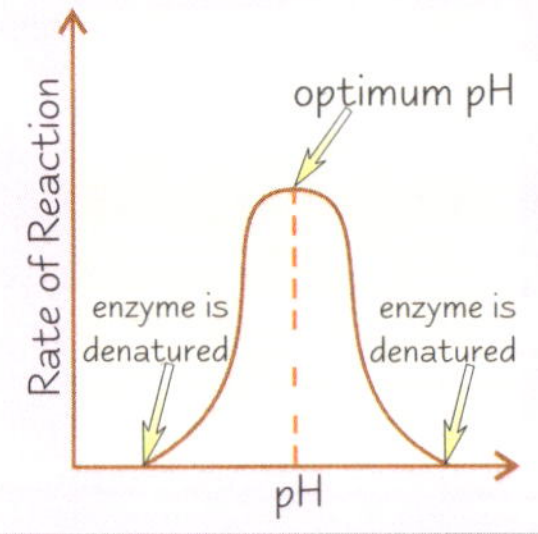

All enzymes have an **optimum pH value**. Most work best at neutral pH 7, but there are exceptions. **Pepsin**, for example, works best at acidic pH 2, which suits it to its role as a stomach enzyme. Above and below the optimum pH, the H+ and OH- ions found in acids and alkalis can mess up the **ionic bonds** that hold the enzyme's tertiary structure in place. This makes the active site change shape, so the enzyme is **denatured**.

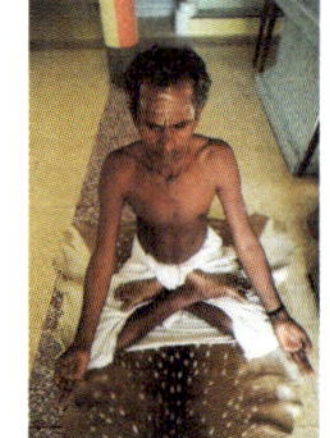

Tony found that revision was possible only under very specific conditions

Enzyme Concentration Affects the Rate of Reaction

1) The **more enzyme molecules** there are in a solution, the more likely a substrate molecule is to **collide** with one. So increasing the concentration of the enzyme increases the rate of reaction.
2) But if the amount of substrate is limited, there comes a point when there's more than enough enzyme to deal with all the available substrate, so adding more enzyme has **no further effect**.

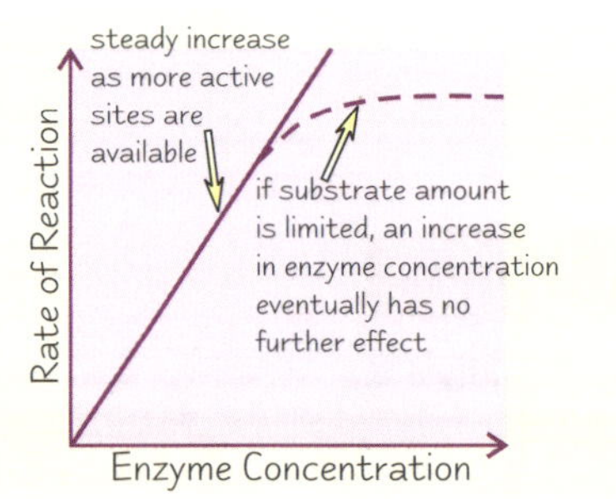

Substrate Concentration Affects the Rate of Reaction Up To a Point

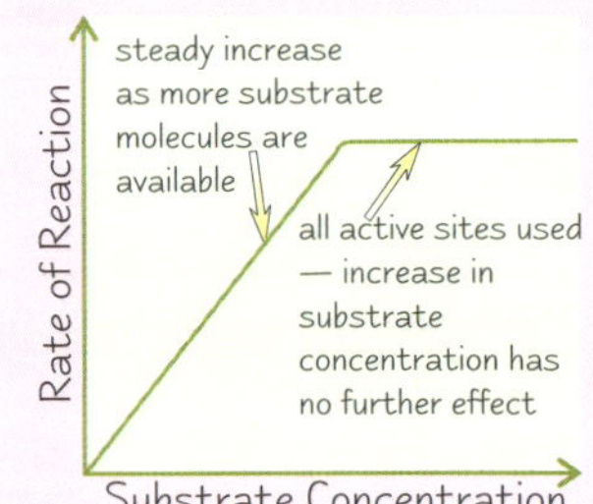

Substrate concentration affects the rate of reaction up to a certain point. The higher the substrate concentration, the faster the reaction, but only up until a **'saturation' point**. After that, there's so many substrate molecules that the enzymes have about as much as they can cope with, and adding more **makes no difference**.

Industrial Production of Enzymes

Enzymes' special properties mean that they are used loads in biotechnology (biotechnology is the use of microorganisms to make useful products). Because enzymes are just so darn important, many of them are mass-produced industrially.

Proteases are Found in Washing Powders

Biological washing powders and liquids contain enzymes to help them **remove stains**.

1) **Proteases** digest (**hydrolyse**) proteins in milk, egg and blood.
2) **Lipases** digest fats.
3) **Cellulases** are sometimes added to get rid of the fluff that dulls the colour of worn fabrics.

Enzymes added to detergents must work well at pH 9-11, and must be thermostable (stable in heat) so they'll work in hot water.

The Food Processing Industry uses Pectin

1) **Pectins** are found in and between plant cell walls. They form **gels** and make the juice thicker in ripe fruit.
2) If powdered **pectinase** enzymes are added to **crushed fruit** the amount of juice that can be extracted increases.
3) Extracted juice can have a **cloudy** appearance — adding pectinase enzymes **clears** it.

Immobilised Enzymes have many Advantages

Immobilised enzymes are attached to or **trapped** in a non-reactive, insoluble material, like a **fibrous polymer mesh.**

1) The enzymes aren't mixed in with the product so it's easy to recover them and **reuse** them. This keeps **production costs** down.
2) The product isn't **contaminated** by the enzyme.
3) Immobilised enzymes are more **pH-** and **heat-stable**.

   E.g. The **dairy industry** uses **immobilised lactase** to convert the lactose in milk into galactose and glucose.

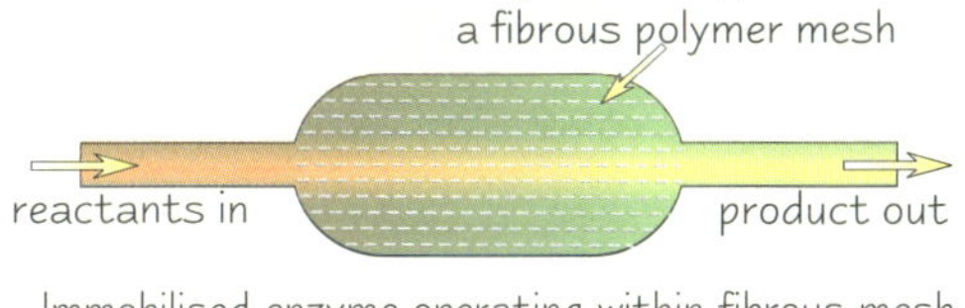

Immobilised enzyme operating within fibrous mesh

Practice Questions

Q1 Why do high temperatures and high or low pH values denature enzymes?

Q2 Explain why increasing the concentration of an enzyme doesn't always increase the rate of reaction.

Q3 Which enzymes do manufacturers add to the detergent in biological washing powders?

Q4 Which type of enzyme is used to make fruit juice look clearer?

Exam Questions

Q1 When doing an experiment on enzymes, explain why it is necessary to control the temperature and pH of the solutions involved. [8 marks]

Q2 a) What is an immobilised enzyme? [2 marks]
b) Explain two advantages of using immobilised enzymes. [2 marks]

Come in Mr. Enzyme, I've been Ex'Pectin you

It's not easy being an enzyme. They're just trying to get on with their jobs, but the whole world seems to be against them sometimes. High temperature, wrong pH, inhibitors — they're all out to get them. And if enzymes didn't exist we'd all have smeary stains down the fronts of our clothes. Sad though it is, make sure you know every word. Learn how different factors affect enzyme activity, and be able to describe how they're used in industry.

Cell Organisation

Woohoo — cells. Riveting, I'm sure. We're all made of cells though, so you can't knock 'em really.

There are **Two Types** of Cell — **Prokaryotic** and **Eukaryotic**

PROKARYOTES	EUKARYOTES
Extremely small cells (0.5-3.0 μm diameter)	Larger cells (20-40 μm diameter)
No nucleus — DNA free in cytoplasm	Nucleus present
Cell wall made of a polysaccharide, but not cellulose or chitin	Cellulose cell wall (in plants and algae) or chitin cell wall (in fungi)
Few organelles	Many organelles
Small ribosomes	Larger ribosomes
Example: *E. coli* bacterium	Example: Human liver cell

Prokaryotic cells are simpler than eukaryotic cells. Prokaryotes include bacteria and blue-green algae. Eukaryotic cells are more complex, and include all animal and plant cells.

Prokaryotic Cells, e.g. Bacteria, are **Small** and have a **Simple Structure**

The only prokaryotes you need to know about for the exam are **bacteria** — in particular rod-shaped bacteria, like *Escherichia coli*. You need to know the **structure** of the cell and what all the different bits inside it (**organelles**) are for.

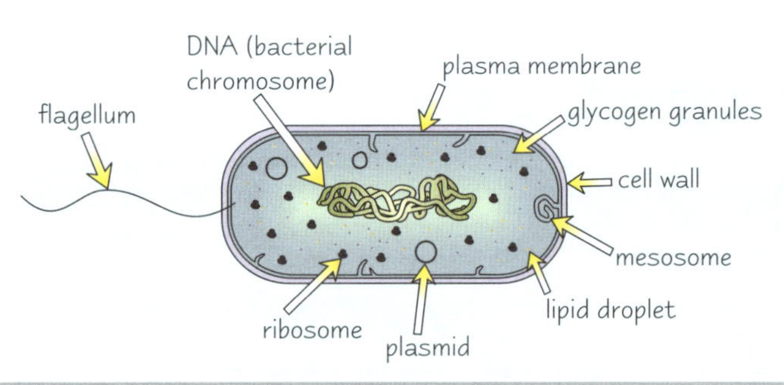

1) The **cell (plasma) membrane** has a very similar structure to that found in eukaryotic cells (see p.24), and has the same function.
2) The **cell wall** is similar to that of a plant in that it supports the cell, but it isn't made from cellulose. It's made of a polymer called **peptidoglycan** instead (don't worry — you don't need to know what that is).
3) The **DNA** of a bacterium isn't enclosed in a nucleus. Instead it floats free in the cytoplasm. Most of the DNA is in one long strand called the **bacterial chromosome**.
4) **Plasmids** are small loops of DNA that aren't part of the bacterial chromosome. Plasmids tend to contain genes for things like **antibiotic resistance**, and can be passed from bacterium to bacterium.
5) **Ribosomes** are the site of protein synthesis. In prokaryotic cells, they are free in the cytoplasm. (See p.13 and p.14 for more details on protein synthesis.)
6) **Glycogen granules** and **lipid droplets** are food storage for the bacterium.
7) The **mesosome** is a bit of the cell membrane that's folded back on itself. This is where **respiration** happens. There are smaller infoldings of the membrane all round the cell, which probably have a similar purpose.
8) The **flagellum** (plural **flagella**) is a long, mobile, hair-like structure that lets the bacterium move.

Most **Eukaryotic** Cells are **Adapted** for **Specific Functions**

Most eukaryotic cells in multicellular organisms are **adapted** to do a particular job. You need to know these examples:

1) **Palisade mesophyll cells** in leaves do most of the photosynthesis. They're **tall** and **narrow**, and can be packed together closely near the surface of the leaf. They contain **many chloroplasts**, many of which are near the top of the cell, so they can absorb as much sunlight as possible.

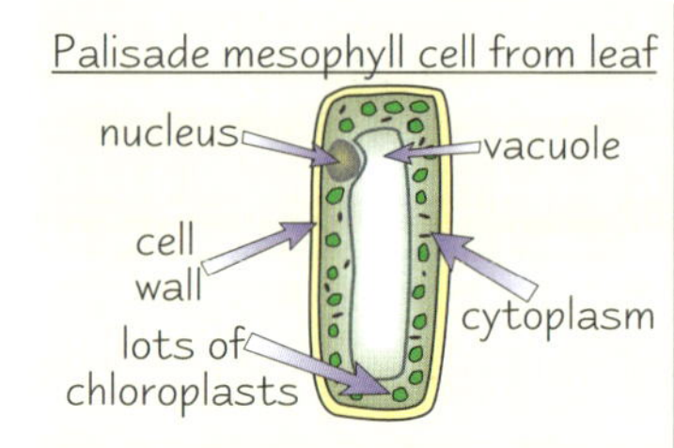

2) **Liver cells** have lots of **mitochondria** and **glycogen granules**. These features aid their function in storing and transferring **energy**. They also have a prominent **Golgi apparatus**, which is useful for **secreting substances** from the cell.

Cell Organisation

Similar Cells are *Organised* into *Tissues*

A single-celled organism performs all its life functions in its one cell. **Multicellular organisms** (like us) are more complicated — different cells do different jobs, so cells have to be **organised** into different groups. Similar cells are grouped together into **tissues**.

A tissue is defined as a group of cells, of **common origin** and similar **structure**, with a particular function.

Tissues aren't always made up of **one** type of cell — many tissues include different types of cell working together.

E.g. Xylem is a plant tissue with two jobs — it **transports water** around the plant, and it **supports** the plant. The cells are mostly **dead and hollow** with no end walls (so they are like tubes) and they have **thick walls** for strength.

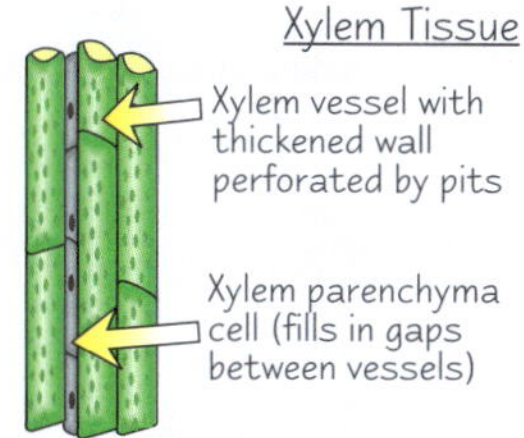

Tissues are *Organised* into *Organs*

An **organ** is a group of tissues that work together to perform a particular function.

The **leaf** is an example of a plant organ.
It's made up of the following tissues:

1) **Lower epidermis** — contains stomata (holes) to let carbon dioxide and oxygen in and out.
2) **Spongy mesophyll** — full of spaces to let gases circulate.
3) **Palisade mesophyll** — most photosynthesis takes place here (see p.20).
4) **Xylem** — carries water to the leaf.
5) **Phloem** — carries sugars away from leaf.
6) **Upper epidermis** — covered in a waterproof waxy cuticle to reduce water loss.

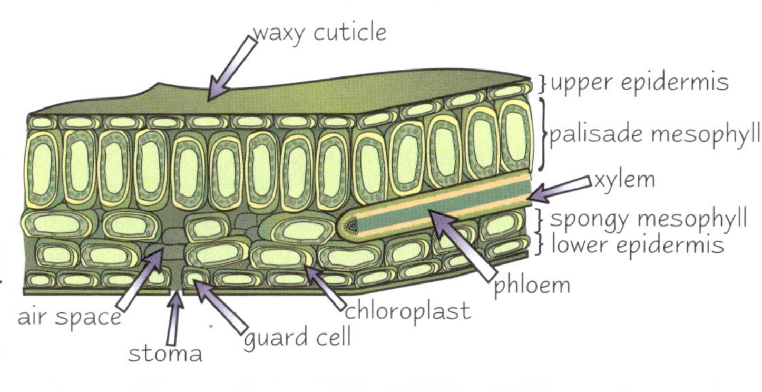

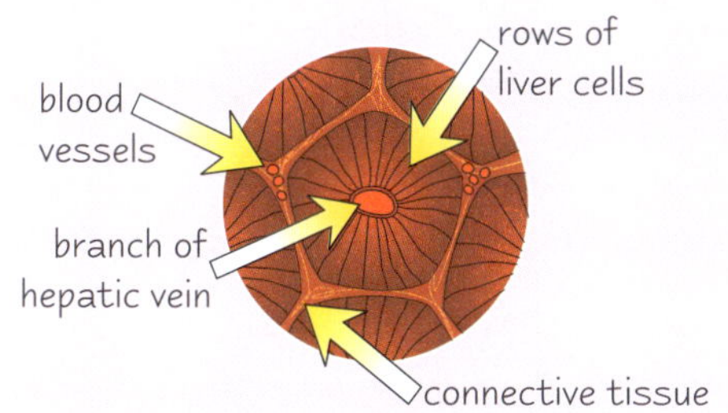

The **liver** is an example of an animal organ.

1) **Liver cells** are the main tissue.
2) There are blood vessels containing blood to provide food and oxygen for the liver cells. Blood is a tissue (yes, really).
3) **Connective tissue** holds the organ together.

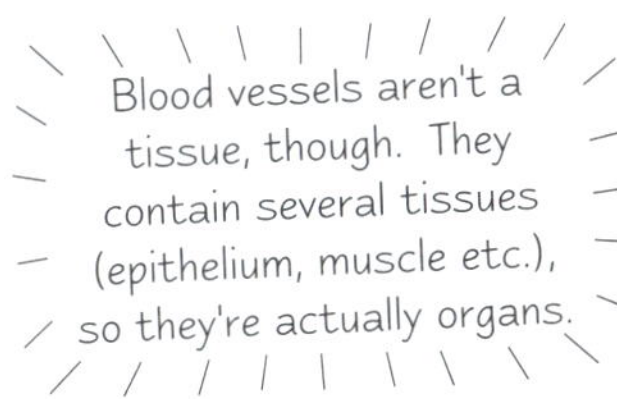

Practice Questions

Q1 Describe, and explain the function of, bacterial flagella.

Q2 What is the definition of a tissue?

Q3 What is the main function of the upper epidermis in a leaf?

Exam Question

Q1 Explain how each of the tissues in a mesophytic leaf is adapted to allow the leaf to perform its function. [12 marks]

Cells — tiny little blobs with important jobs to do...

OK, so there's loads of detail here to learn — but it's fairly easy stuff and will get you many, many beautiful marks in the exam. Make sure you learn the examples of tissues and organs and how cells are adapted to their jobs — examiners love examples. And make sure you know the differences between eukaryotic and prokaryotic cells. And brush your hair — you look like a mess.

Electron and Light Microscopy

You can't get away from microscopes in biology. So you need to learn this page, otherwise all those colourful splodges will remain meaningless for ever more.

Magnification is Size, *Resolution* is Detail

1) **Magnification** is how much bigger the image is than the specimen.
   It's calculated as: $\frac{\text{length of drawing or photograph}}{\text{length of specimen object}}$
2) **Resolution** is how detailed the image is. More specifically, it's how well a microscope distinguishes between two points that are close together. If a microscope lens can't separate two objects, then increasing the magnification won't help.

Light microscopes have a **lower resolution** than electron microscopes. Decreasing the wavelength of the light increases resolution, but even then a light microscope can only distinguish points 0.2 micrometres (μm) apart.

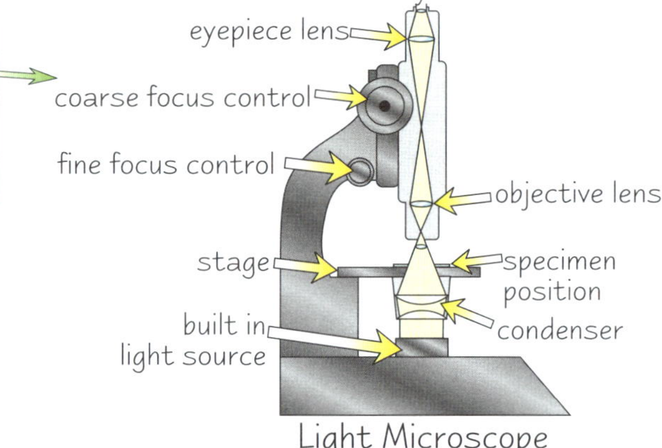

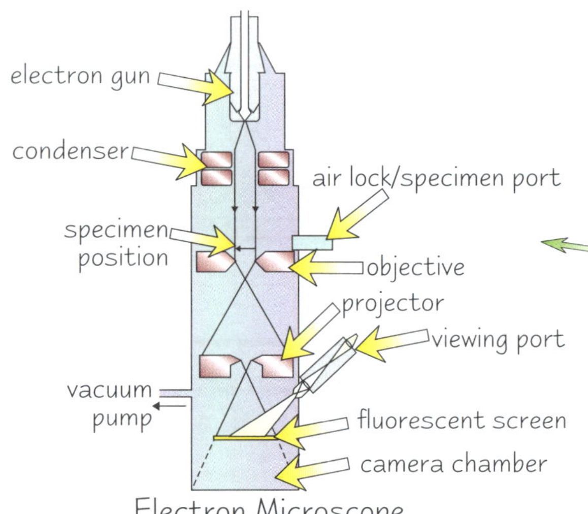

Electron microscopes use **electrons** instead of light to form an image, and focus them with an electromagnet. You can't see electrons, so the image has to be formed on a fluorescent screen. Electrons have a much **shorter wavelength** than light, and can resolve things down to 0.5 nanometres (0.0005 μm) — so electron microscopes provide better resolution and **more detailed images**.

Light Microscopes Show *Cell Structure*

If you just want to see the **general structure of a cell**, then light microscopes are fine. But even the best light microscopes can't see most of the organelles in the cell. You can see the larger organelles, like the nucleus, but none of the internal details.

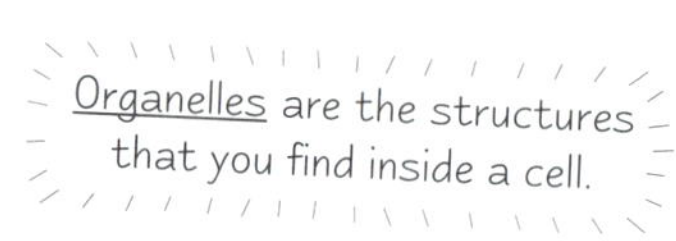

Liver cells seen under a light microscope:

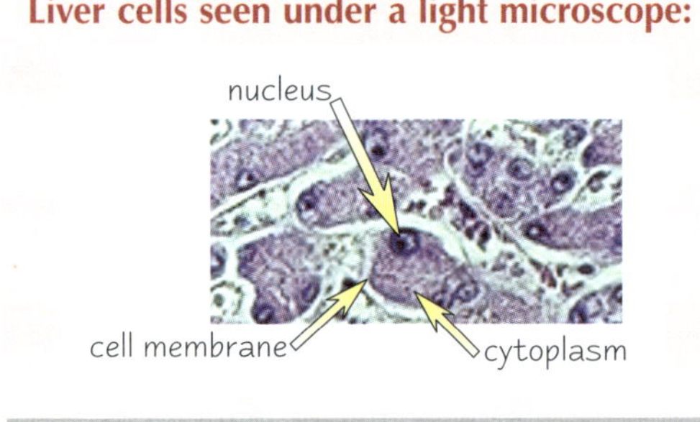

General animal cell as if seen under a light microscope

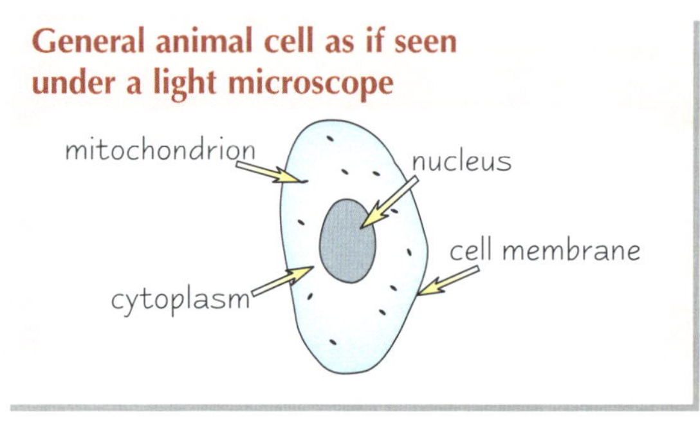

Electron and Light Microscopy

Electron Microscopes Show *Organelles*

There's not much in a cell that an electron microscope can't see. You can see the **organelles** and the **internal structure** of most of them.

Most of what's known about cell structure has been discovered by electron microscope studies. The diagram to the right shows what you can see in an **animal cell** under an electron microscope.

Very pretty indeed.

General animal cell as if seen with an electron microscope

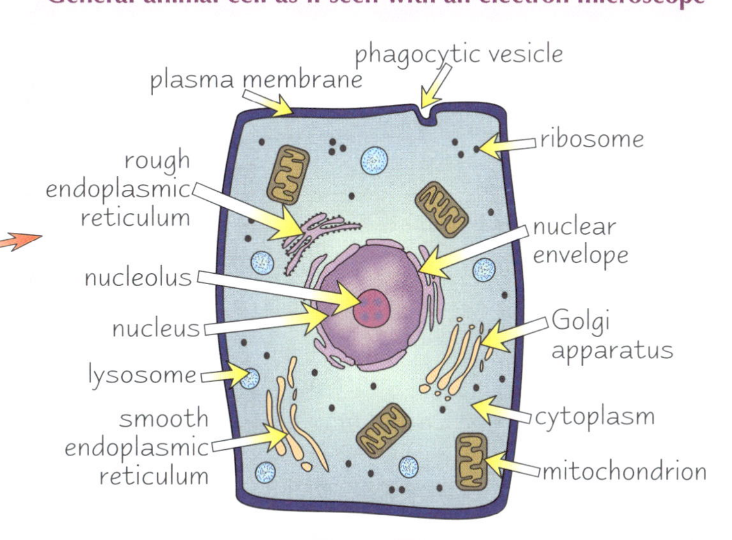

Comparing Light and Electron Microscopy

LIGHT MICROSCOPE	ELECTRON MICROSCOPE
Resolution down to 0.2 µm	Resolution down to 0.5 nm
Living tissue can be examined	Processing kills living cells
Colours can be seen	Natural colours can't be seen
Mobile	Can't be moved around
Relatively cheap	Very expensive

A micrometre (µm) is a thousandth of a millimetre. A nanometre (nm) is a thousandth of a micrometre. That's tiny.

Electron microscopes are ace, but scientists still use light microscopes as well, because they have some advantages.

Practice Questions

Q1 Define the terms 'magnification' and 'resolution' in the context of microscopes.

Q2 Why do electron microscopes have a better resolution than light microscopes?

Q3 What is used to focus an electron microscope?

Q4 Name two organelles that can be seen by light microscopes, and two that can't be seen.

Exam Question

Q1 Explain the advantages and disadvantages of using an electron microscope rather than a light microscope to study cells. [6 marks]

Learn to use a microscope — everything will become clearer...

You need to know the differences between light and electron microscopes, and the advantages and disadvantages of using each one. It's fascinating stuff — almost as fascinating as the time I sat in a room watching paint dry for a very long time. But remember you're revising to get a decent mark in your exam — it'll be worth all the hard work in the end.

Functions of Eukaryotic Organelles

Organelles are all the tiny bits and bobs inside a cell that you can only see in detail with an electron microscope. It's cool to think that all these weird and wonderful things live inside our tiny cells.

Cells Contain **Organelles**

An organelle is a **structure** found inside a cell — each organelle has a **specific function**. Most of these organelles are only found in **eukaryotic cells** (see p.20 for all the prokaryotic organelles you need to know about).

Most organelles are surrounded by membranes, which sometimes causes confusion — don't make the mistake of thinking that a diagram of an organelle is a diagram of a whole cell. They're not cells — they're **parts of** cells, see.

ORGANELLE	DIAGRAM	DESCRIPTION	FUNCTION
Cell wall	plasma membrane; cell wall; cytoplasm	A rigid structure that surrounds **plant cells**. It's made mainly of the carbohydrate **cellulose**.	**Supports** plant cells.
Plasma membrane	plasma membrane; cytoplasm	The membrane found on the surface of **animal cells** and just inside the cell wall of **plant cells**. It's made mainly of **protein** and **lipids** (for more detail, see p.26).	**Regulates the movement** of substances into and out of the cell. It also has **receptor molecules** on it, which allow it to respond to chemicals like hormones.
Nucleus	nuclear membrane; nucleolus; nuclear pore; chromatin	A large organelle surrounded by a **nuclear membrane**, which contains many **pores**. The nucleus contains **chromatin** and a dense structure called the **nucleolus**.	The **chromatin** contains the genetic material (DNA) which **controls the cell's activities**. The pores allow substances (e.g. RNA) to move between the nucleus and the cytoplasm. The **nucleolus** makes **RNA**.
Lysosome		A **round organelle** surrounded by a **membrane**, with no clear internal structure.	Contains **digestive enzymes**. These are kept separate from the cytoplasm by the surrounding membrane, but can be used to **digest invading cells** or to **destroy the cell** when it needs to be replaced.
Ribosome	small subunit; large subunit	A **very small organelle** either floating free in the cytoplasm or attached to rough edoplasmic reticulum.	The **site** where **proteins** are made.
Rough Endoplasmic Reticulum (RER)	ribosome; fluid	A system of membranes enclosing a fluid-filled space. The surface is **covered with ribosomes**.	**Transports proteins** which have been made in the ribosomes.
Smooth Endoplasmic Reticulum		Similar to rough endoplasmic reticulum, but with no **ribosomes**.	**Transports lipids** around the cell.

Functions of Eukaryotic Organelles

ORGANELLE	DIAGRAM	DESCRIPTION	FUNCTION
Golgi Apparatus	vesicle	A group of smooth endoplasmic reticulum consisting of a series of **flattened sacs**. Vesicles are often seen at the edges of the sacs.	It **packages** substances that are produced by the cell, mainly proteins and glycoproteins. It also **makes lysosomes**.
Microtubule		Very small, hollow **cylinders**, made from the substance **tubulin**.	Act as a **guide** for **moving organelles** in the cell.
Mitochondrion	outer membrane inner membrane crista matrix	They are usually oval. They have a **double membrane** — the inner one is folded to form structures called **cristae**. Inside is the **matrix**, which contains enzymes involved in respiration (but, sadly, no Keanu Reeves).	The **site of respiration**, where **ATP** is produced. They are found in large numbers in cells that are very active and require a lot of energy.
Chloroplast	stroma two membranes granum (plural = grana) lamella (plural = lamellae)	A small, **flattened** structure found in **plant cells**. It's surrounded by a **double membrane**, and also has membranes inside called **thylakoid membranes**. These membranes are stacked up in some parts of the chloroplast to form **grana**. Grana are linked together by lamellae — thin, flat pieces of thylakoid membrane.	The **site** where **photosynthesis** takes place. The light-dependent reaction of photosynthesis happens in the **grana**, and the light-independent reaction of photosynthesis happens in the **stroma**.
Centriole		Small, **hollow cylinders**, containing a ring of microtubules, seen in **animal cells** during cell division.	Involved with the **separation of chromosomes** during cell division.

Practice Questions

Q1 Name two organelles found only in plant cells.

Q2 Name two organelles found only in animal cells.

Q3 Explain the differences between rough and smooth endoplasmic reticulum.

Exam Questions

Q1 The presence and number of specific organelles can give an indication of a cell's function.
Give THREE examples of this, naming the organelles concerned and stating their function. [9 marks]

Q2 a) Identify these two organelles seen in an electron micrograph, from the descriptions given below.

(i) A sausage-shaped organelle surrounded by a double membrane.
The inner membrane is folded and projects into the inner space, which is filled with a grainy material.

(ii) A collection of flattened membrane 'bags' arranged roughly parallel to one another.
Small circular structures are seen at the edges of these 'bags'. [2 marks]

b) State the function of the two organelles that you have identified. [2 marks]

Organs and organelles — 'his and hers' biology terms...

Organelle is a very pretty-sounding name for all those blobs. But under a microscope some of them are actually quite fetching — well I think so anyway, but then my mate finds woodlice fetching, so there's no accounting for taste. Anyway, you need to know the names and functions of all the organelles and also what they look like under the microscope.

The Cell Membrane Structure

Two pages all about cell membranes and what they're made of. Try and contain your excitement when you read about the fluid mosaic structure — there have been some nasty cases of extreme over-excitement in the past.

Membranes Control What **Passes Through** Them

Cells and many of the **organelles** inside them are surrounded by **membranes**.
Membranes have a **range of functions**:

1) **Membranes around organelles** divide the cell up into **different compartments** to make the different **functions more efficient** — e.g. the substances needed for **respiration** (like enzymes) are kept together inside **mitochondria**.
2) Membranes control **which substances enter and leave** a cell or organelle.
3) Membranes **recognise** specific chemical substances and other cells.

Cell Membranes have a **'Fluid Mosaic' Structure**

The **structure** of all **membranes** is basically the same.
They are composed of **lipids** (mainly phospholipids), **proteins** and **carbohydrates** (usually attached to proteins or lipids).

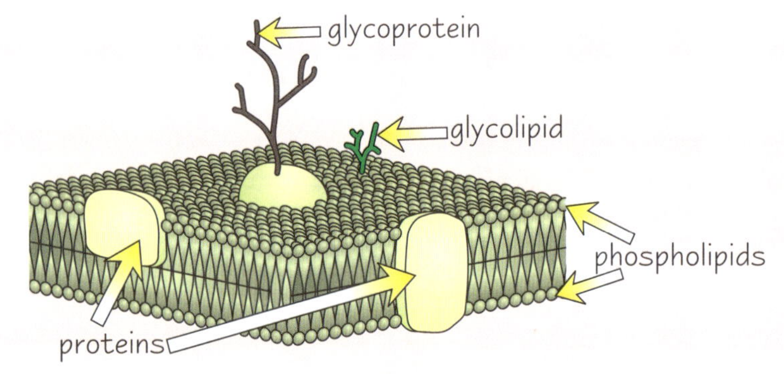

Phospholipid molecules form a continuous, double layer (**bilayer**).

This layer is 'fluid' because the phospholipids are constantly moving. **Protein molecules** are scattered through the layer, like tiles in a **mosaic**.

Phospholipids Can Form **Bilayers**

Phospholipids consist of a **glycerol molecule** plus **two molecules** of **fatty acid** and a **phosphate group** (see p.7).

1) The phosphate / glycerol head is **hydrophilic** — it attracts water.
   The **fatty acid tails** are **hydrophobic** — they repel water.
2) In **aqueous** (**water-based**) **solutions** phospholipids automatically arrange themselves into a **double layer** so that the **hydrophobic tails** pack together **inside the layer** away from the water, and the **hydrophilic heads face outwards** into the aqueous solutions.

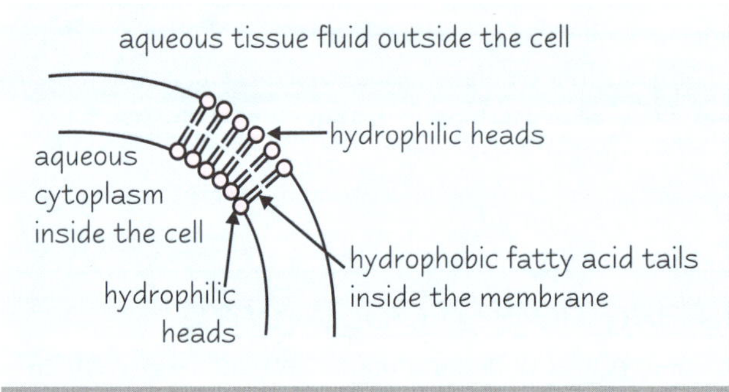

Water soluble molecules (e.g. glucose) can't pass through the fatty, hydrophobic interior of the membrane.

The Cell Membrane Structure

Intrinsic and *Extrinsic* Proteins Have *Different Functions* in the *Membrane*

1) **Intrinsic** proteins **completely span** the membrane from inside to outside.
2) **Extrinsic** proteins only **partly** span the membrane — they're stuck in either the **outer** phospholipid layer or the **inner** phospholipid layer.
3) Intrinsic and extrinsic proteins have **different functions** — usually, intrinsic proteins are for **transport** and extrinsic proteins are **receptors**.

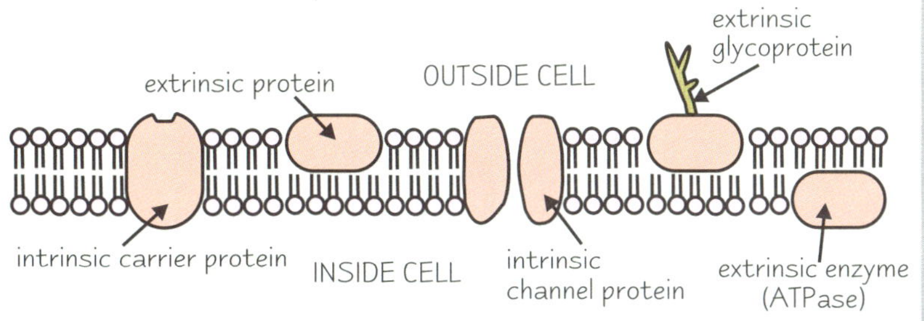

4) **Intrinsic channel proteins** form a tiny **gap** in the membrane to allow water soluble molecules and ions through by diffusion.
5) **Intrinsic carrier proteins** carry water soluble molecules and ions through the membrane by **active transport** and **facilitated diffusion** (see p.30).
6) **Extrinsic** proteins recognise and bind on to **specific molecules** (e.g. hormones).
7) **Enzymes** can be embedded in the inner membrane of a cell or organelle — e.g. ATPase in the inner membrane of mitochondria.
8) Proteins in the membrane also help **strengthen** the membrane. There are **hydrogen bonds** between the proteins and the hydrophilic heads of the phospholipids.

Practice Questions

Q1 Give three functions of cell membranes.

Q2 Which part of a phospholipid molecule is hydrophobic?

Q3 What is the usual function of intrinsic proteins in cell membranes?

Exam Questions

Q1 a) How does a phospholipid differ from a triglyceride? [1 mark]

b) Describe the role of phospholipids in controlling the passage of water soluble molecules through the cell membrane. [2 marks]

Q2 The diagram represents a section of a cell surface membrane.

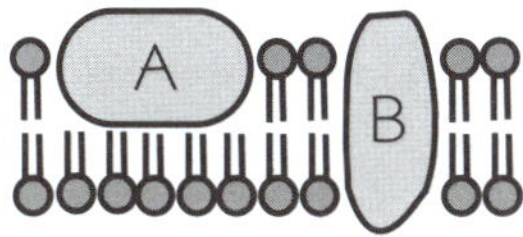

a) Name A and B. [2 marks]

b) Explain how the tertiary structure of a protein molecule allows it to act as a receptor molecule. [2 marks]

*Membranes actually **are** all around...*

The cell membrane is a complex structure — but then it has to be, 'cos it's the line of defence between a cell's contents and all the big bad molecules outside. Don't confuse the cell membrane with the cell wall (found in plant cells). The cell membrane controls what substances enter and leave the cell whereas the cell wall provides structural support.

Transport Across the Cell Membrane

Membranes form a barrier around cells. Substances can be transported across cell membranes in different ways — diffusion, osmosis, facilitated diffusion, active transport and endocytosis/exocytosis. There's four whole pages about them right here.

Diffusion is the *Passive Movement* of *Particles*

1) If there's a **high concentration** of particles (molecules or ions) in one area of a liquid or gas, then these particles will gradually move and **spread out** into areas of **lower concentration**. Eventually, the particles will be **evenly distributed** throughout the liquid or gas. This movement is called diffusion.
2) The path between an area of higher concentration and an area of lower concentration is called the **concentration gradient**. Particles **diffuse faster** when there is a steep concentration gradient (a big difference in concentration between the two areas).
3) Diffusion is described as a **passive process** because **no energy** is needed for it to happen.
4) Diffusion can happen **across cell membranes**, as long as the particles can **move freely** through the membrane. For example, water, oxygen and carbon dioxide molecules are small enough to pass easily through pores in the membrane.

Osmosis is a *Particular Kind of Diffusion*

1) Osmosis is when **water molecules** diffuse through a **partially permeable membrane** from an area of **higher water potential** (i.e. higher concentration of water molecules) to an area of **lower water potential**.
2) A **partially permeable membrane** allows some molecules through it, but not all. Water molecules are small and can diffuse through easily but large solute molecules can't.
3) Water molecules will diffuse **both ways** through the membrane — but the **net movement** will be to the side with a **lower concentration of water molecules**.

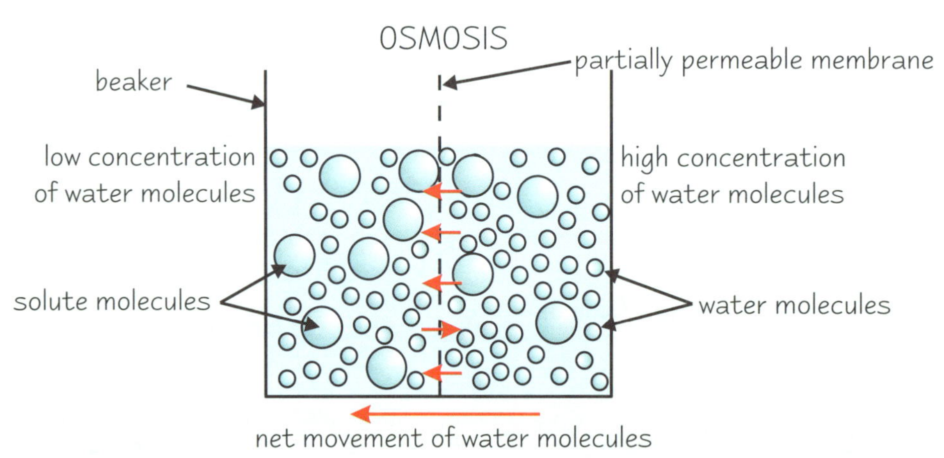

Partially permeable membranes can be useful at sea.

Water Potential is the *Ability* of *Water Molecules to Move*

1) **Water potential** is the potential (likelihood) of water molecules to diffuse out of a solution.

   Water molecules are **more likely** to diffuse out of solutions with a **higher concentration** of water molecules. These solutions have a **high water potential**.

   Water molecules are **less likely** to diffuse out of solutions with a **lower concentration** of water molecules — these have a **low water potential**.

2) Water potential is represented by the symbol ψ. It's measured in **kilopascals** (kPa).
3) **Pure water** has the **highest water potential** and is given the value of **zero kilopascals**. All solutions have a **lower** water potential than pure water, so their water potentials are always **negative**.
4) Water molecules **diffuse** from areas with a **higher water potential** to areas with a **lower water potential**.

Transport Across the Cell Membrane

You can Calculate the Water Potential Inside Cells

The water potential in a cell depends on **two factors**.
You can use these factors to calculate water potential. Clever.

1) **SOLUTE POTENTIAL** (ψ_s)

The amount of **solute molecules** in a solution affects its water potential. Solute molecules form **weak**, **chemical bonds** with water molecules and slow down their movement.
The amount by which solute molecules **reduce** the water potential of a solution is called its **solute potential**.

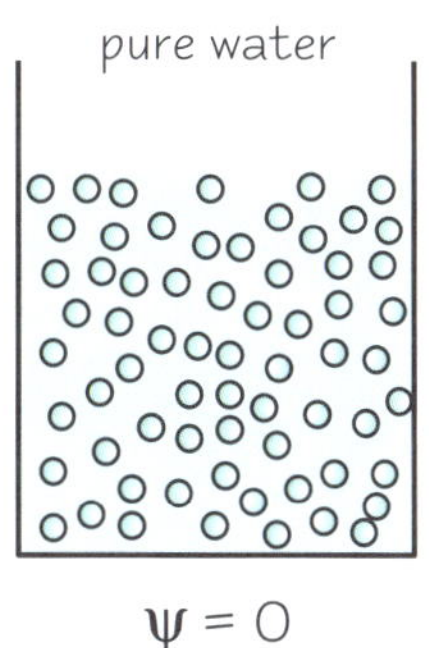

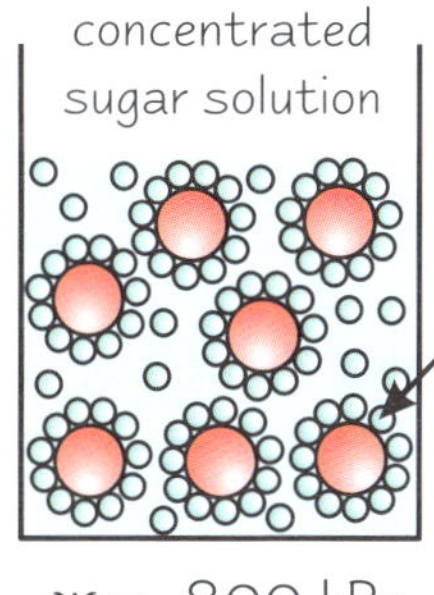

Water molecules bound to solute molecule

The more solute molecules present, the lower (more negative) the water potential (ψ)

2) **PRESSURE POTENTIAL** (ψ_p)

In cells, the water potential is affected by the **cell membrane** (and the cell wall in plants). These **exert pressure inwards** on the cell, in effect squeezing water molecules out the cell. This pressure is called **pressure potential**.

$$\text{water potential } (\psi) = \text{solute potential}(\psi_s) + \text{pressure potential}(\psi_p)$$

Solutions can be Isotonic, Hypotonic or Hypertonic

Isotonic, **hypotonic** and **hypertonic** are terms that describe how the **solute potentials** of solutions **compare** with each other.

Solutions which have the **same solute potential** are **isotonic**.

Their water potentials are therefore the same, so if they're separated by a partially permeable membrane, there is **no net movement** of water between the two.

A **hypotonic** solution has a **lower solute potential**, and therefore a **higher water potential**, than another solution.

There would be a net movement of water **from** the hypotonic solution to the other solution through a partially permeable membrane.

A **hypertonic solution** has a **higher solute potential** and **lower water potential** than another solution.

The net movement of water across a partially permeable membrane would be **into** the hypertonic solution.

Ginantonic solution — my gran's favourite...

*A good way to describe **osmosis** is that it's the diffusion of water molecules through a partially permeable membrane from an area of **higher water potential** to an area of **lower water potential**. Water potential is a tricky idea — but it impresses the examiners, so try and get your head round it.*

Transport Across the Cell Membrane

And there's more...

Cells are Affected by the Water Potential of the Surrounding Solution

Plant and animal cells do different things when they're put into solutions of different concentrations. Flick back a page to see how it's linked to water potential.

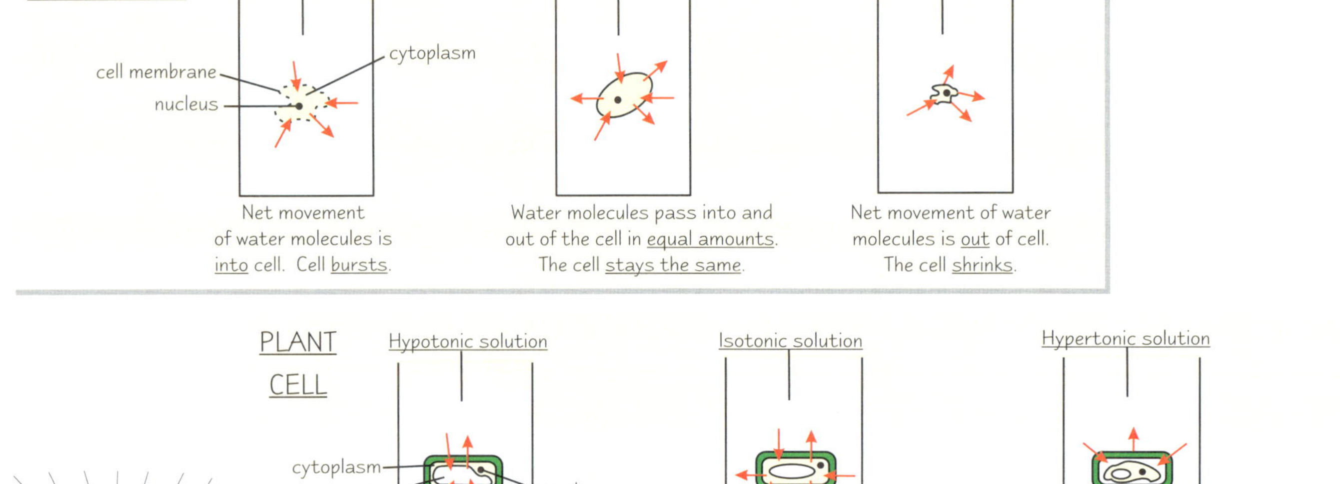

Facilitated Diffusion uses Carrier Proteins and Channel Proteins

Some **larger molecules** (e.g. amino acids, glucose) and **charged atoms** (e.g. sodium ions) can't diffuse through the phospholipid bilayer of the cell membrane themselves. Instead they diffuse through **carrier proteins** or **channel proteins** in the cell membrane. This is called **facilitated diffusion**.

1) Channel proteins form **pores** through the membrane for charged particles to diffuse through.
2) Carrier proteins **change shape** to move large molecules into and out of the cell:

The carrier proteins in the cell membrane have **specific shapes** — so specific carrier proteins can only facilitate the diffusion of specific molecules. Facilitated diffusion can only move particles along a **concentration gradient**, from a higher to a lower concentration. It **doesn't** use any **energy**.

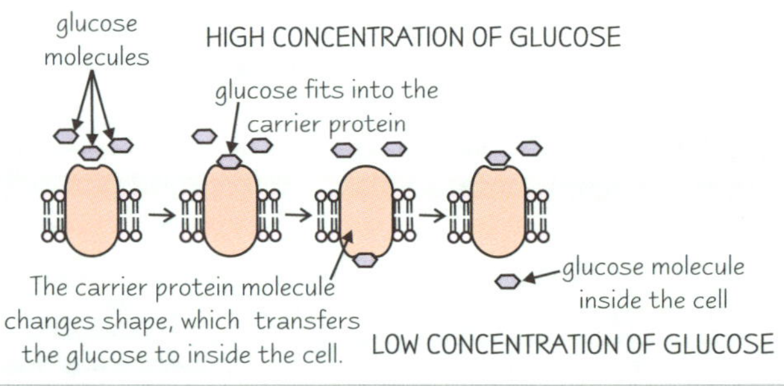

Active Transport Moves Substances Against a Concentration Gradient

1) Active transport uses **energy** to move **molecules** and **ions** across cell membranes, **against** a **concentration gradient**.
2) Molecules attach to **specific carrier proteins** (sometimes called 'pumps') in the **cell membrane**, then **molecules of ATP** (adenosine triphosphate) provide the energy to change the shape of the protein and move the molecules across the membrane.

Transport Across the Cell Membrane

Materials can be Taken into Cells by Endocytosis

Endocytosis is when a cell takes in substances by surrounding them with a section of the cell membrane to form a small vacuole called a **vesicle**.

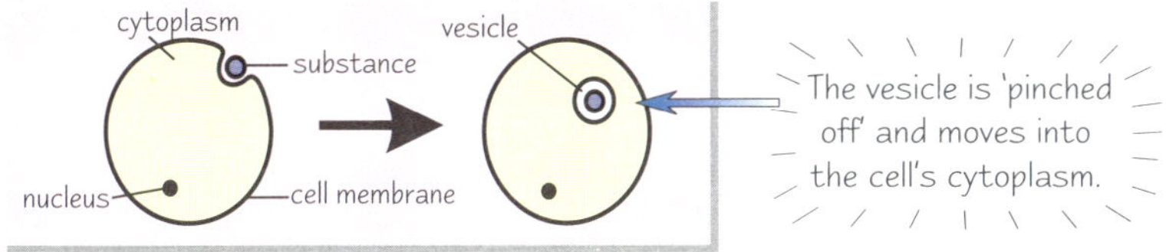

The vesicle is 'pinched off' and moves into the cell's cytoplasm.

There are 2 types of endocytosis:

1) **Phagocytosis** is when solid particles or whole cells are brought into the cell. The contents of the vesicle are **digested** by **enzymes** secreted from **lysosomes**. Molecules and ions then **diffuse** out of the vesicle into the cell's **cytoplasm**.
2) **Pinocytosis** is similar to phagocytosis — but **liquid** is taken into the cell.

Materials can be Removed from Cells by Exocytosis

Materials are **secreted out** of cells by **exocytosis.**

1) **Substances produced by the cell** move through the **endoplasmic reticulum** to the **Golgi body**.
2) **Vesicles** pinch off from the sacs of the Golgi body and move towards the cell membrane. They **merge** with the **cell membrane** and **release** their contents outside of the cell.
3) Digestive enzymes, hormones, mucus and milk are secreted by **exocytosis**.

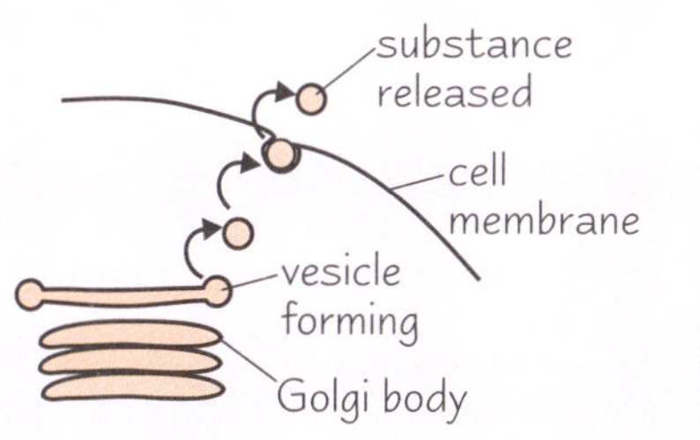

Practice Questions

Q1 What is meant by the term "concentration gradient"?

Q2 What is a partially permeable membrane?

Q3 Describe osmosis.

Q4 Active transport and facilitated diffusion both involve carrier proteins. Which one needs energy?

Q5 What is pinocytosis?

Q6 Give two examples of substances secreted by exocytosis.

Exam Questions

Q1 Write a word equation to define water potential. [2 marks]

Q2 Describe and explain, in terms of water potential, what would happen to an animal cell placed in a hypotonic solution. [3 marks]

Q3 Describe the process of phagocytosis. [5 marks]

A little less conversation, a little more exocytosis, baby...

Phew, the end of a mammoth topic on transport through the cell membrane — so now you can move on and forget it ever happened. Just kidding (I should be doing stand-up, no really) — now you need to go back over it and check you know the details. Learn the differences between similar terms, like hypertonic and hypotonic, and phagocytosis and pinocytosis.

The Cell Cycle and Mitosis

I don't like cell division. There, I've said it. It's unfair of me, because if it wasn't for cell division I'd still only be one cell big. It's all those diagrams that look like worms nailed to bits of string that put me off.

A Chromosome is a Single Strand of DNA Wrapped Round Histone Proteins

Chromosomes are found in the nucleus of eukaryotic cells. Each chromosome is a threadlike structure made up of **one long molecule** of **DNA**. The DNA is wound round **histone proteins**, which support the DNA. It's coiled up very tightly to make a nice, **compact** chromosome.

The DNA molecule is divided into sections. Each section carries the code to make a particular polypeptide (protein). These sections are called genes (see p.11), and each chromosome can carry loads of genes.

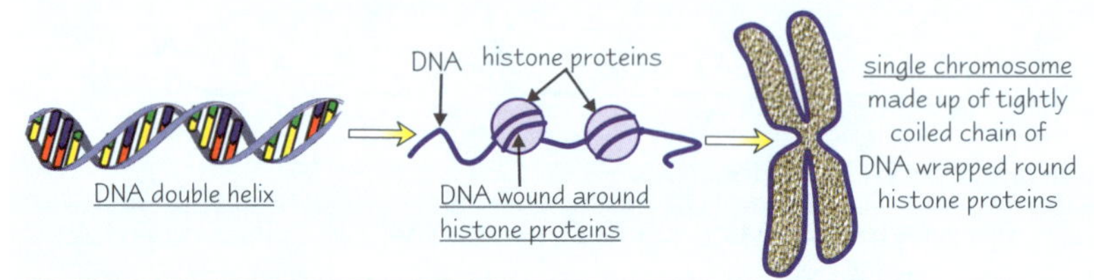

Apart from the X and Y chromosomes in male cells, all the chromosomes of a cell can be matched and put into **pairs**. Pairs of matching chromosomes are called **homologous pairs**. In a homologous pair, both chromosomes are the same shape and size and have the **same genes** in the same location. One chromosome in each pair comes from the female parent, and the other from the male parent.

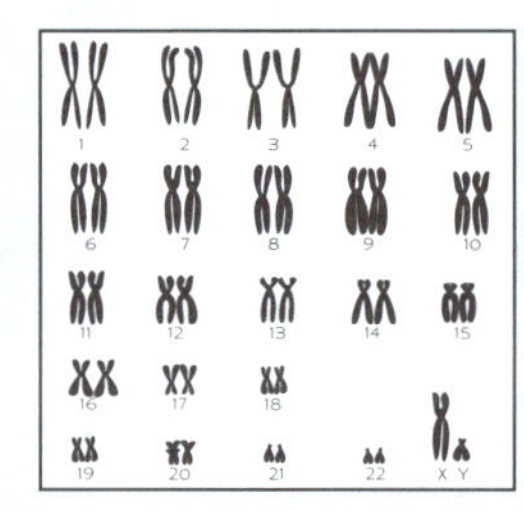

All the cells in an organism contain the **same** chromosomes. Individual cells can then **differentiate** to be specialised for a particular function.

A human cell, e.g. a liver cell, has 46 chromosomes in total — 46 is the **diploid number** (**2n**). Different species have different diploid numbers. For example a palisade leaf cell of a pea plant has a diploid number of 14.

Mitosis is the Cell Division Used in Asexual Reproduction

Cells increase in number by **cell division**. There are two types of cell division — **mitosis** and **meiosis** (see p.54 for more about meiosis). **Mitosis** produces daughter cells that are **genetically identical** to the parent cell. It's used in **asexual reproduction**. It's also needed for the **growth** of multicellular organisms (like us) and for **repairing** damaged tissues.

Cells from multicellular organisms have a clear **cell cycle** that starts when they are produced by cell division, and ends with them dividing themselves to produce more identical cells. The cell cycle consists of a period of cell division called **mitosis** (M phase) and a period in between divisions called **interphase**. Interphase is subdivided into 3 separate growth stages. These are called **G_1**, **S** and **G_2**. Each stage involves specific cell activities:

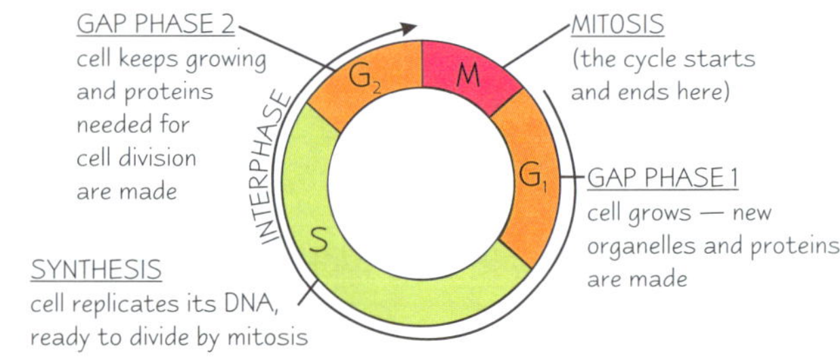

Mitosis has Four Main Stages — Prophase, Metaphase...

Mitosis is really one **continuous process**, but it's described as a series of **division stages** — prophase, metaphase, anaphase and telophase. **Interphase** comes before the division stages — it's when cells grow and replicate their DNA.

1) **Prophase** — The chromosomes **condense**, getting shorter and fatter. Tiny bundles of protein called **centrioles** start moving to opposite ends of the cell, forming a network of protein fibres across it called the **spindle**. The nuclear membrane breaks down and chromosomes lie free in the cytoplasm.

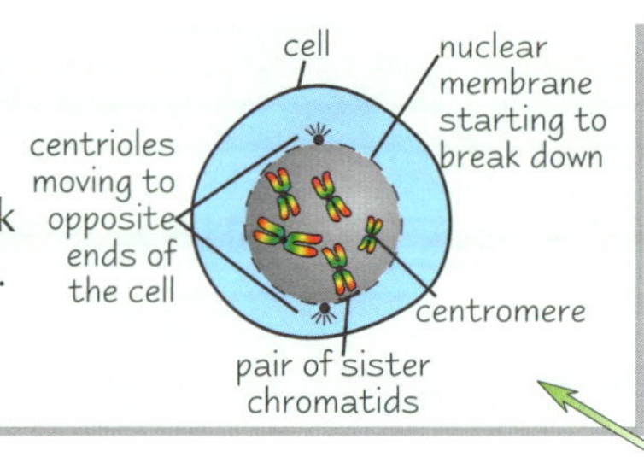

As mitosis begins, the chromosomes are made of two strands joined in the middle by a centromere. The separate strands are called chromatids. There are two strands because each chromosome has already made an identical copy of itself during the synthesis phase of the cell cycle. When mitosis is over, the chromatids end up as one-strand chromosomes in the new daughter cells.

2) **Metaphase** — The chromosomes (each with two chromatids) line up along the middle of the cell and become attached to the spindle by their centromere.

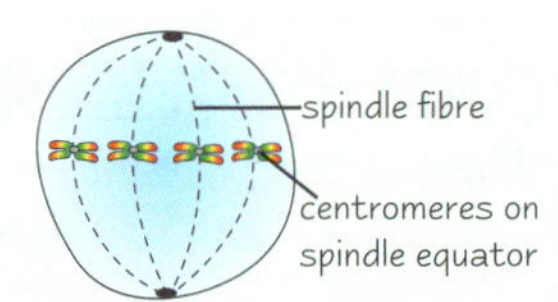

The Cell Cycle and Mitosis

...Anaphase and Telophase

3) **Anaphase** — The centromeres attaching the chromatids to the spindles divide, separating each pair of sister chromatids. The spindles contract, pulling chromatids to opposite poles, centromere first.

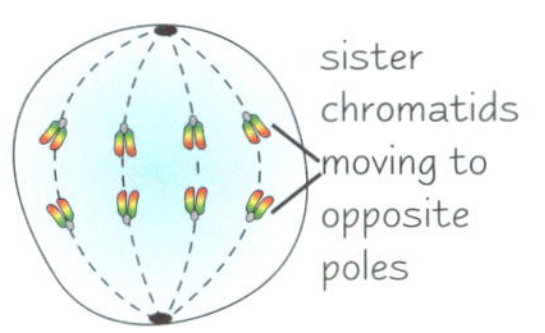

4) **Telophase** — The chromatids reach the **opposite poles** on the spindle. They uncoil and become long and thin again. They're now called **chromosomes** again. A **nuclear membrane** forms around each group of chromosomes, so there are now **two nuclei**. The **cytoplasm divides** and there are now **two daughter cells** which are **identical** to the original cell. Mitosis is finished and each daughter cell starts the **interphase** part of the cell cycle to get ready for the next round of mitosis.

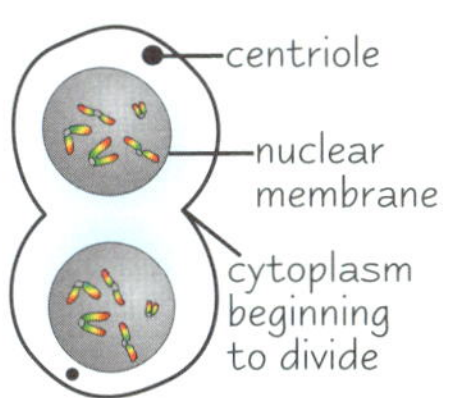

Asexual Reproduction Produces Clones by Mitosis

Asexual reproduction, which happens by **mitosis**, needs only **one** parent. The offspring are normally **genetically identical** to this parent and are called **clones**. Asexual reproduction in plants can be natural or artificial:

1) **Tubers** grow naturally on plants like potatoes, and give clones when planted.
2) Some plant cells are able to divide and form all the cells needed to create a new plant. So humans can **artificially clone** plants by removing parts of them and cultivating them in rooting powder and compost. Artificial cloning from tiny plant specimens (**explants**) is called **micropropagation**:

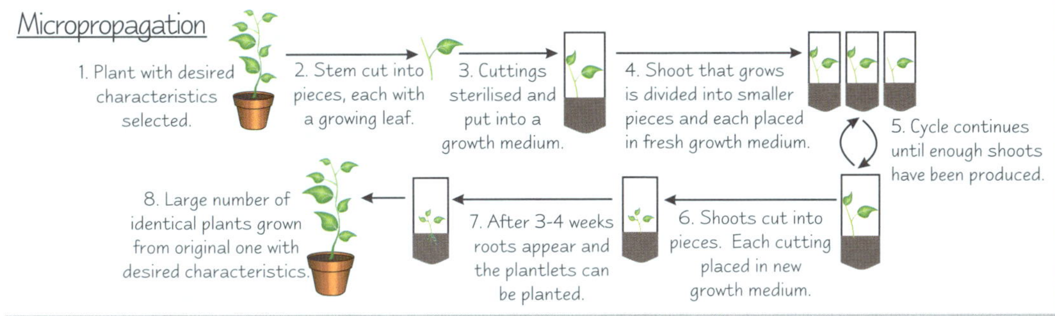

Cloning Animals is Possible but More Complicated

The first mammal clone was **Dolly the sheep**. She was made when the **nucleus** of a **mammary cell** from one ewe was placed into the **ovum** of another ewe, which had had its nucleus removed. This new cell was transferred into the uterus of a different ewe to develop.

Another method is used to clone animals with desirable features, like cows with **high milk yields**. Eggs from the best cow are fertilised in the lab with sperm from the best bull — this is called **in-vitro fertilisation**. The fertilised egg divides, giving a ball of genetically identical cells, which develop into an **embryo**. The young embryo can be **split** into separate cells — each cell grows into a new embryo that's genetically identical to the original one. These embryos are then transplanted into **surrogate** cows to develop.

Practice Questions

Q1 What is meant by asexual reproduction?

Q2 Describe the process of micropropagation.

Exam Question

Q1 During which stages of the cell cycle would the following events take place?

a) DNA replication. [2 marks]

b) Formation of spindle fibres. [2 marks]

Doctor, I'm getting short and fat — don't worry, it's just a phase...

Quite a lot to learn in this topic, but it's all dead important stuff so no slacking. All cells undergo mitosis — from human liver cells to leaf palisade cells. It's how they multiply and how organisms like us grow and develop. Learn those diagrams of mitosis — they usually come up in the exam and you don't want to get your prophase mixed up with your telophase.

Nutrient and Gaseous Exchange

These pages are about surface area. Basically, things get into and out of organisms across exchange surfaces, which are usually cell membranes. There needs to be a big enough area of surface (see what I did there?) for this to happen.

Organisms Need to **Exchange Materials** with the **Environment**

Every organism, whatever its size, needs to exchange materials with its environment.

1) Cells need to take in **oxygen** (for aerobic respiration) and **nutrients**.
2) They also need to excrete **waste products** like CO_2 and urea.

How easy it is to do this depends on the **surface to volume ratio** of the organism.

Smaller Animals have Higher **Surface Area : Volume Ratios**

A mouse has a bigger surface area **relative to its volume** than a hippo. This can be hard to imagine, but you can prove it mathematically. Imagine these animals as a series of cubes:

The hippo could be represented by a block with sides of 2 cm × 4 cm × 4 cm.

Its **volume** is 2 × 4 × 4 = **32 cm³**

Its **surface area** is 2 × 4 × 4 = 32 cm² (top and bottom surfaces of block)
+ 4 × 2 × 4 = 32 cm² (four sides of the block)

Total surface area = **64 cm²**

So the hippo has a **surface area : volume ratio** of 64 : 32 or **2 : 1**.

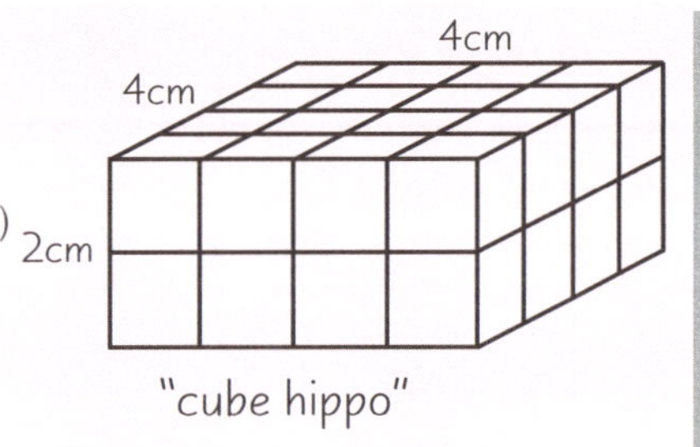

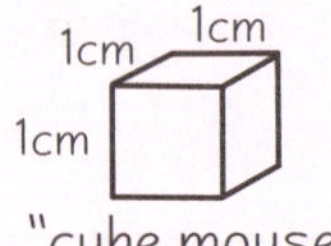

Compare this to a mouse cube measuring 1 cm × 1 cm × 1 cm

Its **volume** is 1 x 1 x 1 = **1 cm³**

Its **surface area** is 1 x 1 x 6 = **6 cm²**

So the mouse has a **surface area : volume ratio** of 6 : 1

The cube mouse's surface area is six times its volume.

The cube hippo's surface area is only twice its volume.

Smaller animals have a **bigger surface area** compared to their **volume**.

Most Multicellular Organisms need **Specialised Exchange Mechanisms**

Organisms with different surface area : volume ratios exchange materials in different ways...

1) Microscopic one-celled organisms like **protozoa** have very **high** surface area : volume ratios, so they can exchange materials with the environment over their whole surface by **diffusion**.
2) Larger multicellular organisms have **lower** surface area : volume ratios. They need **organs** with **big surface areas** and **specialised cells** for exchange (e.g. lungs — see p.36 and p.37). Many have also developed specialised **transport systems** like blood, to carry gases, nutrients and wastes to and from inner cells.

Surface area is also important for **body temperature**. Animals release heat energy when their **cells respire** and lose it to the **environment**. So **small animals** with high S.A. : volume ratios, like mice, have to use lots of energy just **keeping warm**, while **big animals** with low S.A. : volume ratios, like elephants, are more likely to **overheat**.

Elephants have a low S.A. : volume ratio, but their big, flat ears have a very small volume and a big surface area. Elephants use their ears for heat exchange, to help heat escape from the body quickly.

Gaseous Exchange in Flowering Plants

Plants *Exchange Gases* at the *Surface* of the *Mesophyll Cells*

Two processes in plants involve gas exchange:

1) **Respiration takes in oxygen** and **releases carbon dioxide** — just like in animals.
2) **Photosynthesis** uses energy from sunlight to **convert carbon dioxide** and **water into glucose and oxygen**. The light energy is absorbed by chlorophyll.

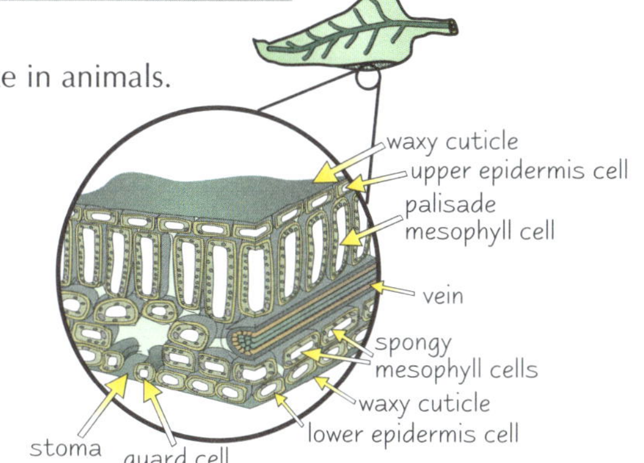

The main gas exchange surface is the **surface of the mesophyll cells** in the leaf. This is well adapted for its function — it's **moist**, and there's a **large surface area**.

The mesophyll cells are inside the leaf. Gases pass back and forth from the outside through special pores in the **epidermis** called **stomata** (singular = stoma). The stomata can open to allow exchange of gases, and close if the plant is losing too much water.

Guard Cells Help Control when *Stomata* Open and Close

1) Each stoma is surrounded by **guard cells**.
2) The stoma opens when the guard cells **increase in turgidity**.
3) They become turgid by absorbing water by **osmosis**. When this happens, each guard cell changes shape because its **inner cell wall** is **thicker** than its outer one. This opens the pore, as the diagram shows.

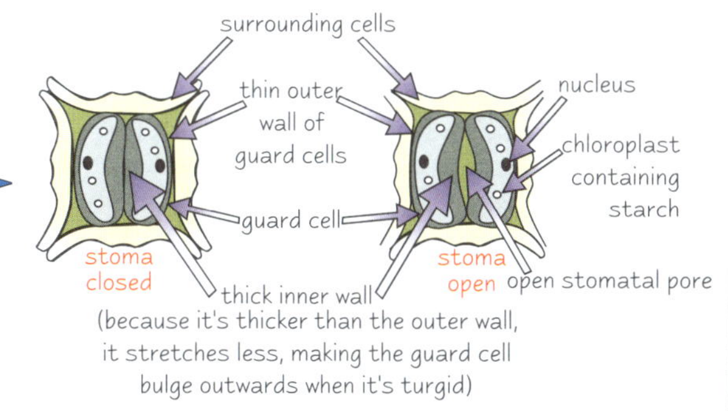

Stomata are generally **open** during the **day** so that CO_2 can enter the plant for **photosynthesis** (which needs **light**). They are usually **closed at night** to **save water**, because CO_2 is no longer needed (photosynthesis can't happen in the dark). The **plant controls** when stomata open and close:

When **light shines** on the guard cells, or when there are **low CO_2 concentrations** in the leaf, **two things happen**:

1) A **potassium pump** is triggered, which actively pumps **K^+ ions** into the guard cells from **surrounding cells**.
2) Starch, stored in the **chloroplasts** of guard cells is converted to **malate ions**.
Starch is insoluble and **doesn't** affect osmosis. **Malate** is soluble and **does affect osmosis**.

This **increases the concentrations** of potassium and malate ions in the guard cells, which **reduces the water potential** of the guard cells (see p.29). This means water is drawn in by osmosis, which **opens the stomata**.

The opposite happens in the dark — the potassium pump stops and malate is converted back to starch. This reduces the concentration of ions in guard cells, so water potential increases. Water leaves the guard cells by osmosis and the stomata close for the night. Sweet dreams.

Practice Questions

Q1 What is the main gas exchange surface in plants?

Q2 What ions are important in the opening and closing mechanism of stomata?

Exam Questions

Q1 Explain why a specialised gas exchange system is required in humans. [4 marks]

Q2 Stomata open and close due to internal changes in the guard cells.
Explain the mechanism that causes these changes. [6 marks]

I don't get it — what's stomata with me...

Ahem — sorry. Moving swiftly on... I know you've just got to the end of a page — but it would be a pretty smart idea to have another look at diffusion and osmosis (p.28). It's all that stuff about transport across membranes. Not the most thrilling prospect I realise, but it'll help these pages make more sense.

Gas Exchange in Humans

Just a couple of pages on breathing here — all pretty straightforward...

Lungs are Specialised Organs for Breathing

Mammals, e.g. humans, exchange oxygen and carbon dioxide through their **lungs**. The lungs have special **features** that make them well-adapted to **breathing**:

Cartilage — rings of strong but bendy cartilage keep the **trachea** open.

Goblet cells — produce **mucus** to trap inhaled dust and other particles.

Cilia — **hairs** on the cells that line the trachea, bronchi and bronchioles. They **move** to push the mucus with trapped particles **upwards**, away from the lungs.

Smooth muscle — round the bronchi and bronchioles. **Involuntary** muscle contractions narrow the airways.

Elastic fibres — between the alveoli. Stretch the lungs when we breathe in and recoil when we breathe out to help push air out.

Pleural membrane — protective lining on the lungs.

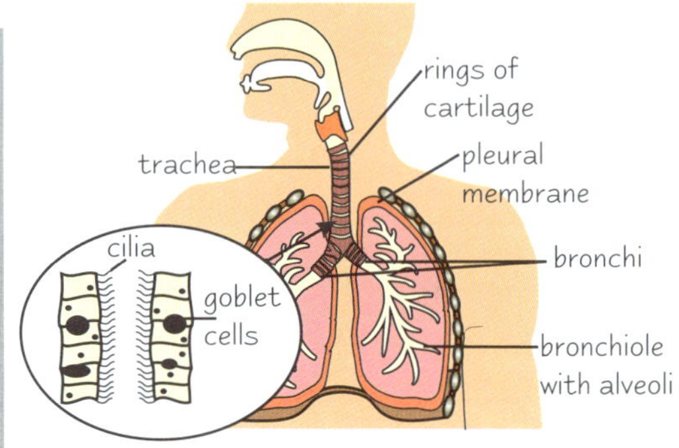

TV is the Volume of Air Taken into the Lungs in One Breath

Tidal Volume (TV) is the volume of air in each breath in or out — usually about **0.4 dm³** per breath when you're resting.

Vital Capacity (VC) is the maximum volume of air that can be breathed out after a deep breath.

dm³ is short for decimeters cubed.

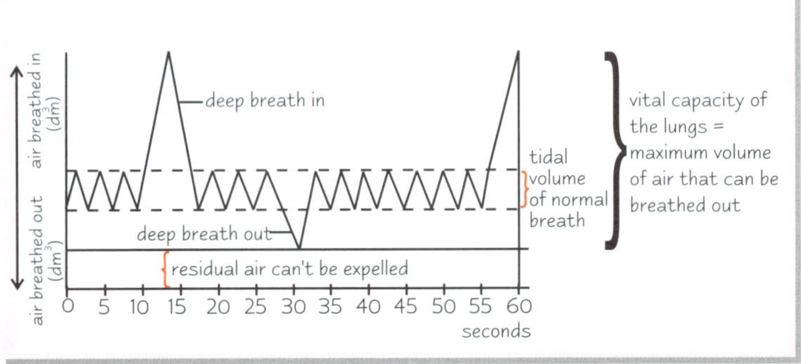

The Medulla Oblongata in the Brain Controls Rate of Breathing

Your **breathing rate** changes according to how much **physical activity** you're doing. When you're exercising you use more **energy**. Your body needs to do more **aerobic respiration** to release this energy, so it needs more **oxygen**.

The **ventilation cycle** is the cycle of breathing in and out. It involves **inspiratory** and **expiratory** centres in the **medulla oblongata** (an area of the brain) and **stretch receptors** in the lungs.

The Ventilation Cycle:

1) The medulla's **inspiratory centre** sends nerve impulses to the **intercostal** (rib) and **diaphragm** muscles to make them **contract**. It also sends nerve impulses to the medulla to **inhibit** the **expiratory centre**.
2) Air enters the lungs due to the **pressure difference** between the lungs and the air outside.
3) The **lungs inflate**. This stimulates **stretch receptors**, which send nerve impulses back to the **medulla** to **inhibit** the **inspiratory centre**.
4) Now the expiratory centre (no longer inhibited) sends nerve impulses to the muscles to relax and the **lungs deflate**, expelling air. This causes the **stretch receptors** to become **inactive**, so the inspiratory centre is no longer inhibited and the cycle starts again.
5) The lungs are surrounded by **pleural membranes** which are lubricated so the ribcage can move smoothly during inhalation and exhalation.
6) This ventilation cycle happens **automatically** without you having to think about it.

Gas Exchange in Humans

In Humans, ***Gaseous Exchange*** *Happens in the* ***Alveoli***

Lungs contain millions of microscopic air sacs called **alveoli**, which are responsible for gas exchange. They're so tiny, but so important — size ain't everything.

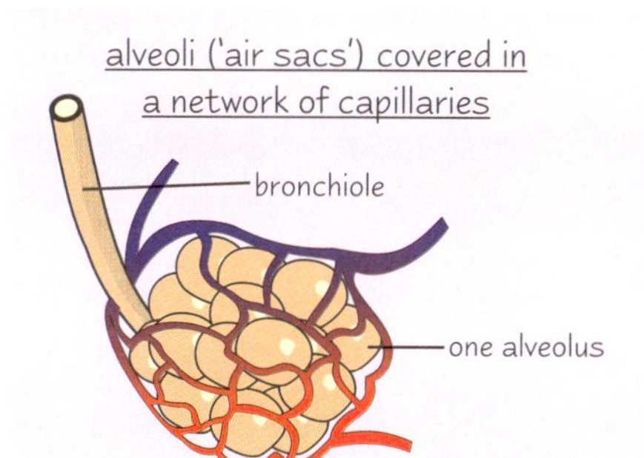

1) The huge number of alveoli means a **big surface area** for exchanging oxygen and carbon dioxide.
2) $\mathbf{O_2}$ diffuses **out of** alveoli, across the **alveolar epithelium** (the single layer of cells lining the alveoli) and the **capillary endothelium** (the single layer of cells of the capillary wall), and into **haemoglobin** in the **blood**.
3) $\mathbf{CO_2}$ diffuses **into** the alveoli from the blood, and is breathed out through the lungs, up the trachea, and out of the mouth and nose.
4) The alveoli secrete liquid called **surfactant**. This stops the alveoli collapsing by lowering the surface tension of the water layer lining the alveoli.

Alveoli have **adaptations** that make them a really good surface for gas exchange. They have the following features, which all **speed up** the **rate of diffusion**:

- **thin exchange surfaces** (the alveolar epithelium cells are very thin);
- **short diffusion pathways** (the alveolar epithelium layer is only one cell thick);
- **a large surface area to volume ratio**;
- **a steep concentration gradient** between the alveoli and the capillaries surrounding them.

Chemoreceptors *Detect* ***Chemical Changes*** *in the* ***Blood***

1) During exercise, CO_2 levels rise and this decreases the **pH** of the blood.
2) **Chemoreceptors** are sensitive to these chemical changes in the blood. They're found in the **medulla oblongata**, in **aortic bodies** (in the aorta), and in **carotid bodies** (in the carotid arteries carrying blood to the head).
3) If the chemoreceptors **detect** a **decrease** in the **pH** of the blood, they send a **signal** to the **medulla** to send more frequent nerve impulses to the intercostal muscles and diaphragm. This **increases** the **rate** of **breathing** and the **depth** of breathing.
4) This allows **gaseous exchange** to **speed up**. CO_2 levels drop and the demand for extra O_2 by the muscles is met.

Practice Questions

Q1 Give five features of the lungs which make them well-adapted for breathing.

Q2 Explain the role of stretch receptors in the ventilation cycle.

Q3 Write down four ways in which the alveoli are adapted for gas exchange.

Q4 What structures detect changes in pH in the blood?

Q5 Where is the rate of breathing controlled?

Exam Question

Q1 How would you expect tidal volume to change when an athlete runs a long race?
Explain your answer. [3 marks]

Alveoli — useful things...always make me think of pasta...

I may have been a bit optimistic when I said this was straightforward. OK, I was lying. There are loads of terms to learn, and it's tempting just to read them and hope for the best. But the safest way is to close the book and try to write them out, along with their meanings. Keep doing this until the cows come home and the fat lady sings — or, until you know them all.

Digestion and Absorption

Time for a bit of blood and guts — well, actually, mainly just the guts. The human digestive system demands respect — it's an amazing bit of kit that has to deal with whatever you trough. It's no coincidence that 'gutsy' means well 'ard.

*The **Human Gut** is **Adapted** to its **Function***

There are two stages in human digestion:

1) **Mechanical breakdown** of large pieces of food into small pieces.
2) **Chemical breakdown** of large molecules into small molecules.

The human gut (**alimentary canal**) is composed of different parts, each with a specific job to do:

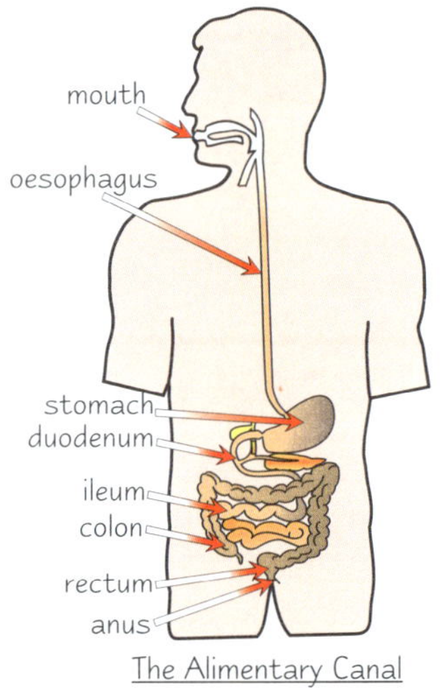

The Alimentary Canal

1) **Mouth** — **Mastication** (chewing) of food by teeth **mechanically** breaks up food so there's a **larger surface** area for enzymes to work on. Mixing food with saliva (water, **amylase** and mucus) partly digests food so it can be swallowed easily. The amylase **hydrolyses** starch into maltose.

2) **Oesophagus** — A tube that takes food from the mouth to the stomach using waves of muscle contractions called **peristalsis**. Mucus is secreted from glandular tissue in the walls, to lubricate the food's passage downwards.

3) **Stomach** — The stomach walls produce **gastric juice**, which consists of hydrochloric acid (HCl), **pepsin** (an enzyme) and mucus. Pepsin is an **endopeptidase** — it hydrolyses peptide bonds in the **middle** of polypeptide molecules (proteins), breaking them down into smaller polypeptide chains. It only works in **acidic conditions**, which are provided by the HCl. Peristalsis in the stomach turns food into an acidic fluid, called **chyme**.

4) **Duodenum** (small intestine) — Contains **alkaline bile** and **pancreatic juice**, which neutralise chyme and break food down into small, soluble molecules:

- Bile is produced in the **liver** and stored in the **gall bladder**, then enters the duodenum through the bile duct. It **emulsifies** lipids into small droplets, which speeds up hydrolysis of lipids by **pancreatic lipase**.
- Pancreatic juice contains **digestive enzymes**:

  Lipase — Hydrolyses **lipids** into fatty acids and glycerol.
  Amylase — Hydrolyses **starch** into maltose.
  Trypsin — An **endopeptidase** that hydrolyses polypeptides into smaller polypeptides.
  Exopeptidases — Hydrolyse peptide bonds found at the **end** of polypeptide chains, giving **free amino acids**.

- **Intestinal juice** is produced by the gut wall. It contains more digestive enzymes — more lipases, **maltase** (hydrolyses maltose into glucose), more exopeptidases and also **dipeptidases** that hydrolyse dipeptides into amino acids.

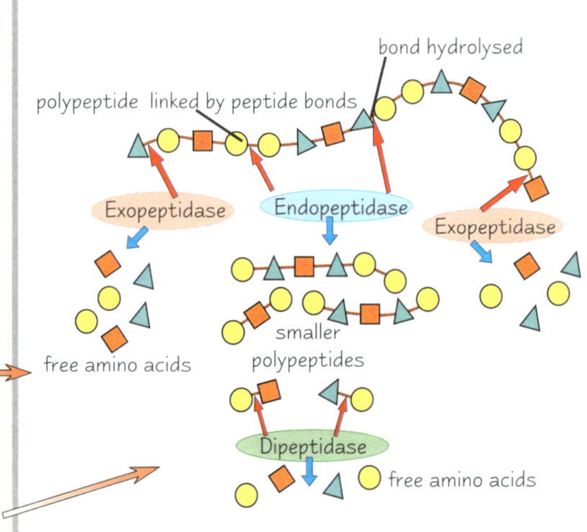

5) **Ileum** (small intestine) — The small, soluble molecules of digested food (glucose, amino acids, fatty acids and glycerol) are absorbed through the **microvilli** lining the gut wall. Absorption is through diffusion, facilitated diffusion and active transport (see p.28 - 30):

Method of Absorption	Affected Molecules
Diffusion	Fatty acids, glycerol, water
Facilitated Diffusion	Some glucose, some amino acids
Active Transport	Mineral ions, some glucose, some amino acids

6) **Colon** (large intestine) — **Water** is **absorbed** by the body and waste is pushed along towards the rectum.
7) **Rectum** — Stores **faeces** until they're expelled through the **anus**.

Digestion and Absorption

The Gut Wall Consists of Four Layers of Tissue

The gut wall has the same **general structure** all the way through the human gut:

1) The **mucosa** (inner lining) lubricates the passage of food with **mucus**. This prevents **autodigestion** (enzymes attacking the gut wall). It's lined with **surface epithelium** cells.
2) The **submucosa** contains capillary beds and nerve fibres.
3) The **circular and longitudinal muscles** control the shape and movement of the gut.
4) The **serosa** contains tough tissue, which provides protection from friction against other organs.

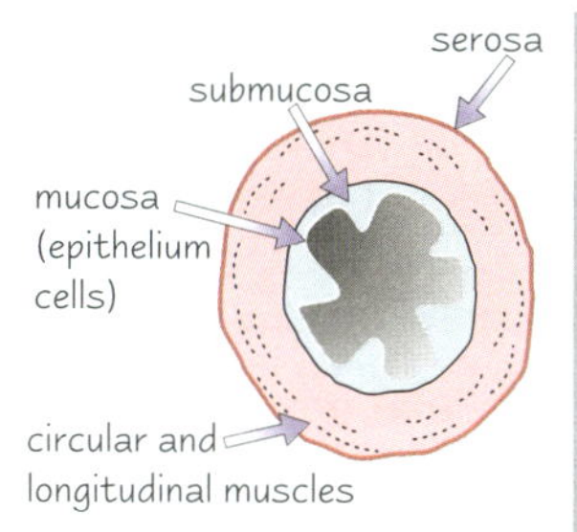

The gut wall in different regions of the alimentary canal has **special features** so it can carry out specific functions:

The gut wall in the **oesophagus** uses its circular and longitudinal muscles to perform **peristalsis**. These muscles work as an **antagonistic pair** — as one contracts, the other relaxes. Food is pushed along in front of the contractions.

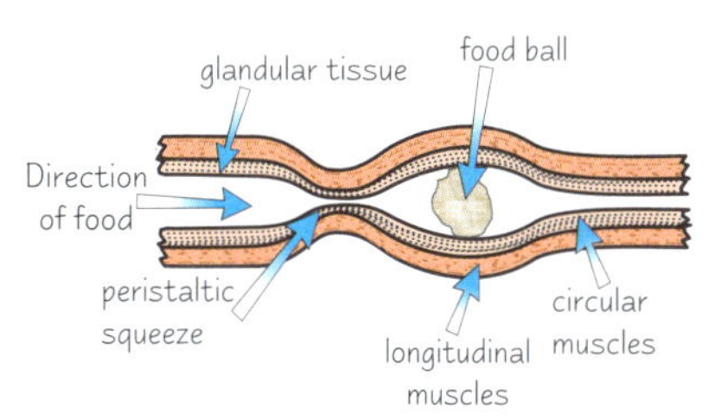

The mucosa in the **ileum** has **villi** and **microvilli** to **increase the surface area** for absorbing the products of digestion. It also has other adaptations for effective absorption:

- It consists of a **single layer** of epithelial cells, so there's a **short diffusion pathway**.
- A **moist lining** helps substances **dissolve** so they can pass through cell membranes.
- The **capillary bed** takes away **absorbed molecules** so the diffusion gradient is maintained.
- **Lymph vessels** take away absorbed **glycerol** and **fatty acids** to join the lymphatic system (you don't need to know about that in any detail). This maintains the diffusion gradient.
- **Carrier proteins** in epithelial cell membranes allow **facilitated diffusion**.
- The epithelial cells contain lots of **mitochondria** to make ATP, needed for **active transport**.

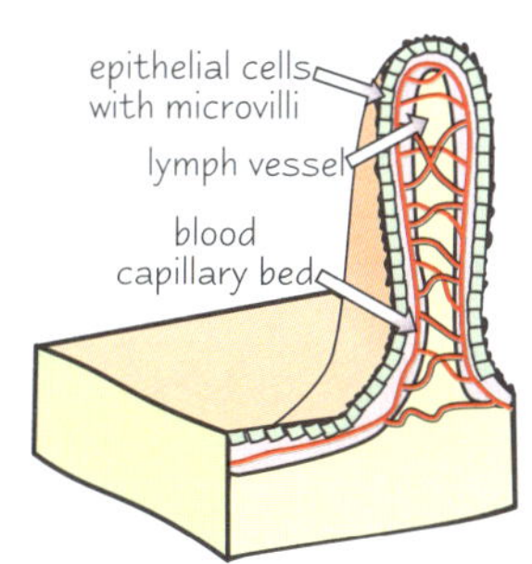

Practice Questions

Q1 What is saliva made of?

Q2 What is peristalsis?

Q3 What do exopeptidases do?

Q4 What is the difference in function between the duodenum and the ileum?

Q5 Draw the layers of the gut wall, label them and describe their functions.

Exam Questions

Q1 Explain how protein is broken down in the digestive system. [4 marks]

Q2 Describe what role the ileum plays in the human digestive system. [3 marks]

My mum told me mastication would make me go blind…

Quite a bit to learn here — it's understandable, 'cos digestion is a complicated process with many stages. The key thing is to learn what each bit of the gut does and how it's adapted to do this function. Make sure you know the difference between endo- and exopeptidases. Then eat some food, and revise its process down the body, not lingering on the last stage though.

Transport in Animals and Plants

After I've told you why organisms have transport systems, I'll tell you about xylem and phloem — the two transport tissues in plants. The xylem and phloem fall in love and they get married.

Multicellular Organisms need Transport Systems

All cells need energy — most cells get energy via **aerobic respiration**. The raw materials for this are **glucose** and **oxygen**, so organisms have to make sure they can deliver enough of these to all their cells. In single-celled creatures, these materials can **diffuse directly** into the cell across the cell surface membrane. The diffusion rate is quick because of the small distances the substances have to travel.

In **multicellular** organisms, diffusion across the outer membrane is too slow for their needs. This is because:

1) some cells are **deep within the organism**;
2) they have a **low surface area to volume ratio** (see p.34);
3) animals have a **high metabolic rate**, which means they respire quickly, so they need a **constant supply** of glucose and oxygen;
4) they have a **tough outer surface**.

The bulk movement of substances around a multicellular organism using a transport system is called mass flow.

Transport systems carry raw materials from specialised **exchange organs** to the cells. In mammals this is the **circulatory system**, which uses **blood** to carry glucose and oxygen around the body. It also carries hormones, antibodies (to fight disease) and waste like CO_2.

Xylem and Phloem are the Transport Systems in Plants

Xylem and **phloem** are found throughout the plant — they **transport materials** to all parts of the plant. **Phloem tissue** transports **solutes** (dissolved substances — mainly sucrose) round plants. The **xylem tissue** transports **water** and **mineral ions** up through plants.

Xylem is Made Up of Four Types of Cell

Vessel elements and **tracheids** are the cells that do the actual **transporting**. They are surrounded by extra cells for **packing** (**parenchyma** cells) and **support** (**fibre** cells).

Vessels — These are very long, **tube-like** structures formed from cells (**vessel elements**) joined end to end. There are **no end walls** on these cells, making an **uninterrupted tube** that allows water to pass through easily. The vessels are **dead**, containing **no cytoplasm**. Their walls are thickened with a woody substance called **lignin**, which helps with **support**. The amount of lignin increases as the cell gets older. Substances get into and out of the vessels through small **pits** in the walls, where there is **no lignin**.

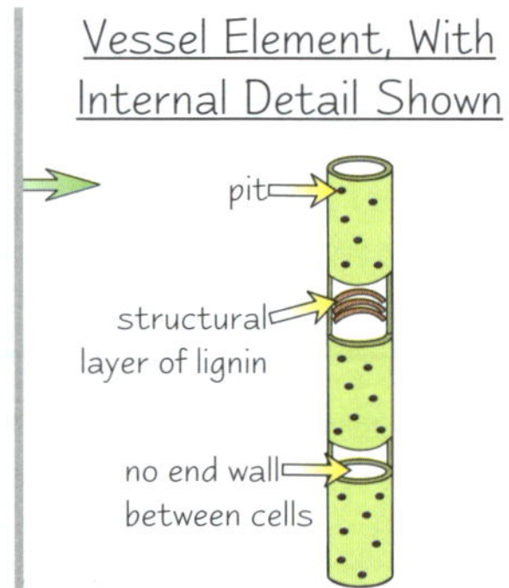

Tracheids — Like vessels, these are long, **lignified** cells with no cytoplasm. They don't form continuous tubes like vessels do — instead they **overlap** so that water can pass through. Where cells overlap the wall contains extra pits, so substances can pass easily between cells. These cells are a **more primitive** form of xylem — plants usually have more vessels than tracheids.

Tracheid Cells

structural layer of lignin

pit

perforated end wall with lots of pits, where cells overlap

cells overlap like this

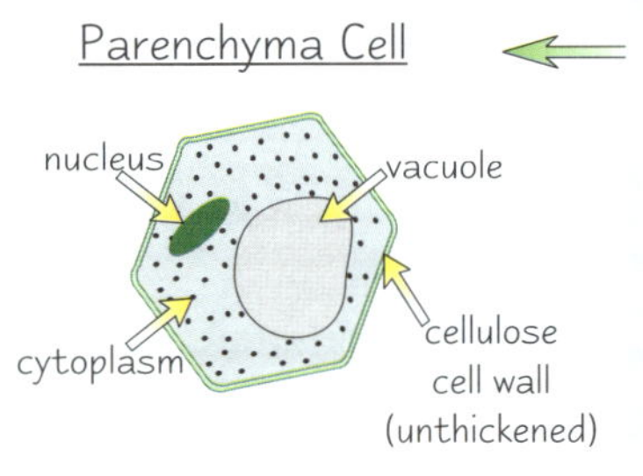

Parenchyma Cells — These are the only **living cells** in the xylem. Their cell walls aren't lignified or thickened. They surround the vessels and tracheids. Their functions include food storage and **radial transport** of food and water. **Radial transport** is movement **across** the stem from the outside to the middle, rather than up and down it.

Fibre Cells — These are small, dead cells with very **thick, lignified** cell walls. Their walls are so thick because the fibres are only used for **support and strength**. They **don't** carry water, so they don't need a wide lumen (the space in the middle).

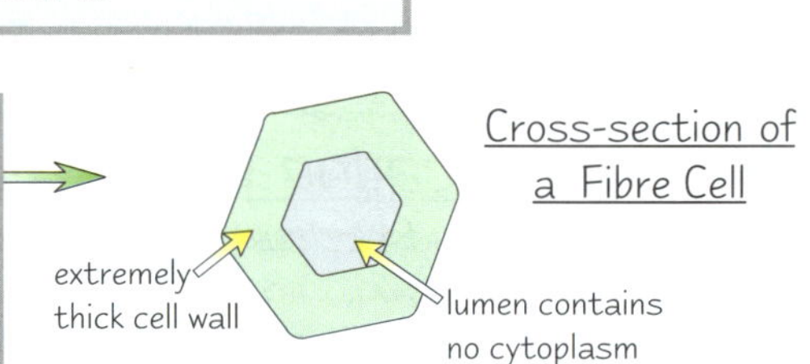

Transport in Flowering Plants

There are Two Main Cell Types in Phloem

Like xylem, **phloem** is formed from cells arranged in **tubes**, and these cells are modified for transport.

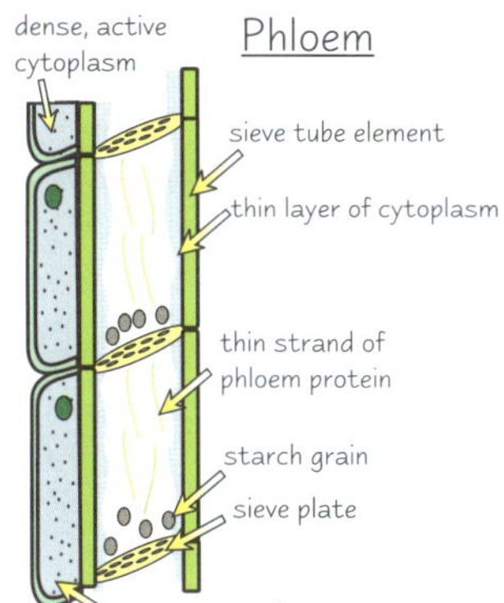

Sieve tube elements — These are **living cells** that **transport solutes** through the plant. They are joined end-to-end to form **sieve tubes**. The 'sieve' parts are the end walls, which have lots of holes in them. Unusually for living cells, sieve tube elements have **no nucleus** and only a **very thin layer of cytoplasm** without many organelles. The cytoplasm of adjacent cells is connected through the holes in the sieve plates.

Companion cells — The lack of a nucleus and other organelles in sieve tube elements means that they would have difficulty surviving on their own. So there is a companion cell for every sieve tube element. The companion cell has a very **dense and active cytoplasm**, and it seems to carry out the living functions for both itself and its sieve cell. Both are formed from a single cell during the development of the phloem.

Phloem tissue also contains **phloem fibres** and **phloem parenchyma**, but these aren't so important for transport.

Plants Need to Absorb Mineral Ions

Plants absorb **dissolved mineral ions** from the soil. This can happen by **diffusion**, but diffusion **isn't** a selective process — and plants often **only** need **certain ions** and not others. So instead, ions needed by a plant are absorbed by **active transport** (see p.30). This means a plant can absorb more of a certain ion, even if it already has a **high concentration** of them **inside** its cells — so it doesn't have to rely on there being a **diffusion gradient** into the plant. It also means that the plant can **pump out** any ions it **doesn't need**.

Mineral ions travel up the plant with the water, in the **xylem**. Scientists know this from using **radioactive tracers**. This uses radioactive forms of ions, as radioactivity can be detected and the scientist will know where the ion has gone.

Amino Acids and Sugars are 'Translocated'

1) Sugars (mostly **sucrose**) are transported from the **leaves** (where they're made during photosynthesis) to **actively growing regions**, or to **storage sites**.
2) Amino acids are made in the **root tips** (where nitrogen is absorbed), and are carried to **growing areas** in the plant to **make proteins**.

The way that they're transported in the phloem isn't known exactly, but it is known that it's an active process, needing energy from respiration.

Translocation is the **movement of dissolved organic substances** (mainly sucrose) to **where they're needed** in the plant. Experiments show that it happens in the **phloem**. Translocation moves substances from '**sources**' to '**sinks**'. The **source** of a substance is **where it's made** (so it's in **high concentration** there). The **sink** is the area where it's **used up** (so it's in **low concentration** there). For example, the source for sugars is the **leaves**, and the sinks are the other parts of the plant, especially the **food storage organs** and **growing points** in roots, stems and leaves.

Enzymes maintain a concentration gradient from the phloem to the sink by modifying the organic substances at the sink.

Practice Questions

Q1 Give two reasons why multicellular organisms need transport systems.

Q2 State two functions of xylem cells in plants.

Q3 What are the two main types of cells found in phloem tissue?

Exam Questions

Q1 The xylem tissue contains several different types of cell. Describe the structure and function of each type. [13 marks]

Q2 Describe the structure and arrangement of sieve tube elements. [5 marks]

Do you, xylem, take phloem, to be your lawful wedded tissue type...

Please don't write the part about them getting married in the exam. I just put that in to make sure you were paying attention. It's vital your mind doesn't wander on this page, because the structure and functions of some of these cell types are quite similar. It can be easy to get mixed up if you haven't learnt it properly, so take the time now to sort out which cell type does what.

Movement of Water

Water enters a plant through its roots and eventually, if it's not used, exits via the leaves. "Ah-ha," I hear you say, "— but how does it flow upwards, against gravity?" Well that, my friends, is a mystery that's about to be revealed.

Water Enters a Plant through its Root Hair Cells

Water has to get from the **soil**, across the **root** and into the **xylem**, which takes it up the plant. The bit of the root that absorbs water is covered in **root hairs**. This increases its surface area and speeds up water uptake. Once it's absorbed, the water has to get through two root tissues, the **cortex** and the **endodermis**, to reach the xylem.

Water always moves from areas of **higher water potential** to areas of **lower water potential** — it goes down a **water potential gradient**. The **soil** around roots has a **high water potential** (i.e. there's lots of water there) and **leaves** have a **low water potential** (because water constantly **evaporates** from them). This creates a water potential gradient that keeps water moving through the plant in the right direction, **from roots to leaves**.

There are Three Routes Water can Take through the Root

Water can travel through the roots into the xylem by three different paths:

1) The **apoplast pathway** — goes through the **non-living** parts of the root — the **cell walls**. The walls are very absorbent and water can simply diffuse through them, as well as passing through the spaces between them.
2) The **symplast pathway** — goes through the **living** cytoplasm of the cells. The **cytoplasm** of neighbouring cells connects through **plasmodesmata** (small gaps in the cell walls).
3) The **vacuolar pathway** — water travels from the vacuole of one cell into the vacuole of the next by **osmosis**.

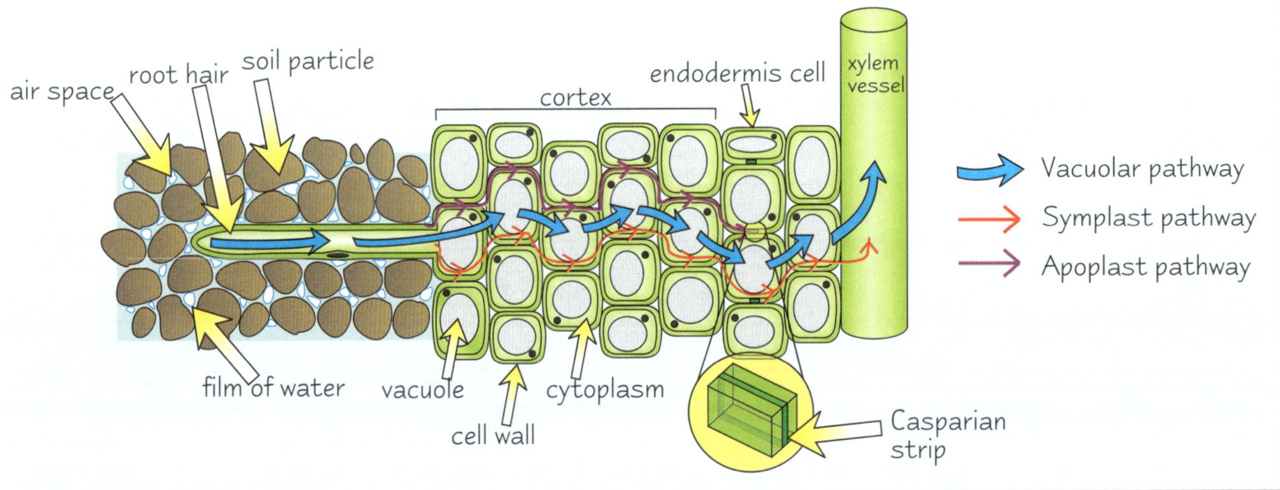

All three pathways are used, but the main one is the **apoplast pathway** because it provides the **least resistance**. When the water gets to the **endodermis** cells, though, the apoplast pathway is blocked by a **waxy strip** in the cell walls, called the **Casparian strip**, which the water can't penetrate. Now the water has to take one of the other pathways. This is useful, because it means the water has to go through a **cell membrane**. Cell membranes are able to control whether or not substances in the water get through (see p.26). Once past this barrier, the water moves into the **xylem**.

Movement of Water

*Water Moves **Up** a Plant **Against** the Force of **Gravity***

The **cohesion-tension theory** explains how water moves up plants from roots to leaves, against the force of gravity.

1) Water evaporates from the leaves at the 'top' of the xylem (transpiration).
2) This creates a **suction** ('tension'), which pulls more water into the leaf.
3) Water molecules **stick together** ('cohesion' — see p.2), so when some are pulled into the leaf others follow.
4) This means the whole **column** of water in the xylem, from the leaves down to the roots, moves upwards.

1) water lost by transpiration
2) water drawn into leaves by suction (tension)
3) water pulled up the stem by cohesion
4) more water enters stem from roots

It's like what happens when you suck at the top of a drinking straw and the liquid moves up it.

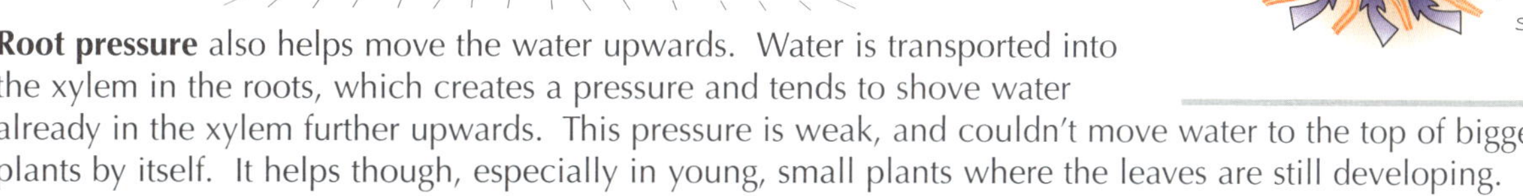

Root pressure also helps move the water upwards. Water is transported into the xylem in the roots, which creates a pressure and tends to shove water already in the xylem further upwards. This pressure is weak, and couldn't move water to the top of bigger plants by itself. It helps though, especially in young, small plants where the leaves are still developing.

***Transpiration** is **Loss of Water** from a Plant's Surface*

The **transpiration stream** is a constant flow of water from the roots to the leaves through the xylem vessels. **Transpiration** is a side effect of photosynthesis. **Photosynthesis** uses up carbon dioxide and water and **produces oxygen** and **glucose**. Water **evaporates** from the moist cell walls and accumulates in the spaces between cells in the leaf. Then it diffuses out of the **stoma**ta when they open. This happens because there is a **diffusion gradient** — there's more water inside the leaf than in the air outside.

Four** Main Factors Affect **Transpiration Rate

The factors below affect transpiration rate. Temperature, humidity and wind alter the **diffusion gradient**, but **light** is a bit different:

1) **Light** — Transpiration happens mainly when the stomata are open. In the dark the stomata usually close, so there's little transpiration.
2) **Temperature** — Diffusion involves the movement of molecules. Increasing the temperature speeds this movement up. So as temperature rises, so does transpiration rate.
3) **Humidity** — If the air around the plant is humid, the **diffusion gradient** between the leaf and the air is reduced. This slows transpiration down.
4) **Wind** — Lots of air movement blows away water molecules from around the stomata. This **increases** the diffusion gradient, which increases the rate of transpiration.

So the rate of transpiration is fastest when it's light, warm, dry and windy.

Practice Questions

Q1 Name the three pathways by which water travels across the root.

Q2 What is the Casparian strip and in which root tissue would you find it?

Q3 Name the theory which explains how water travels up a plant against the force of gravity.

Q4 What four factors affect transpiration rate?

Exam Questions

Q1 Describe the three routes water can take through the roots. [6 marks]

Q2 Explain why movement of water in the xylem stops if the leaves of a plant are removed. [4 marks]

So many routes through the roots…

Lots of impressive biological words on this page, to amaze your friends and confound your enemies. Go through the page again, and whenever you see a word like plasmodesmata, just stop and check you know exactly what it means. (Personally, I think they should just call them cell wall gaps, but nobody ever listens to me.)

Circulation, the Heart and the Cardiac Cycle

*Blood pumps **continually** round your body. It's happening right now as you read this rather long and dull sentence that I'm writing to keep you reading without stopping to try and highlight the **unceasingness** of blood flow. And relax.*

Mammals** Have a **Closed Double Circulation

Animals need circulatory systems to transport respiratory gases, products of digestion, metabolic wastes and hormones round the body. Mammals have a **closed circulatory system** which has two routes around the body:

1) **Systemic circulation**
(heart → body → heart)
Oxygenated blood travels to the body cells and **deoxygenated** blood returns to the heart.

2) **Pulmonary circulation**
(heart → lungs → heart)
Deoxygenated blood travels to the lungs and **oxygenated** blood returns to the heart.

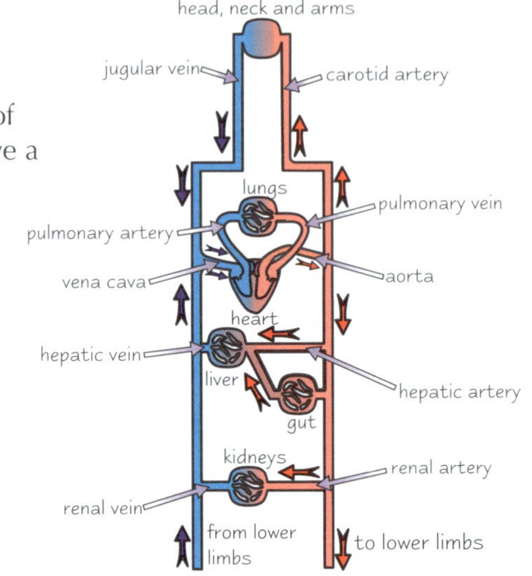

Mass Flow of Blood — Human Closed Circulatory System

*The **Heart** Consists of **Two Muscular Pumps***

The diagram below shows the **internal structure** of the heart. The **right side** pumps **deoxygenated blood** to the **lungs** and the **left side** pumps **oxygenated blood** to the **whole body**. NB — the **left and right side** are **reversed** on the diagram, 'cos it's the left and right of the person that the heart belongs to.

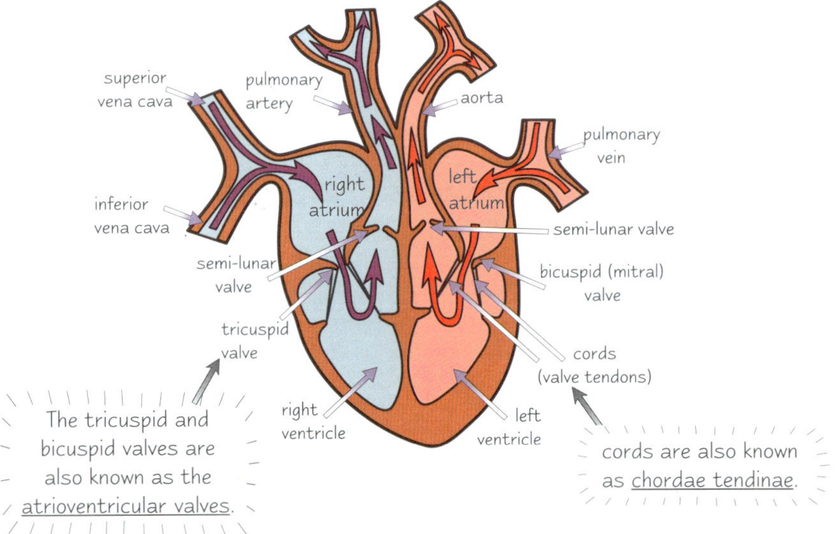

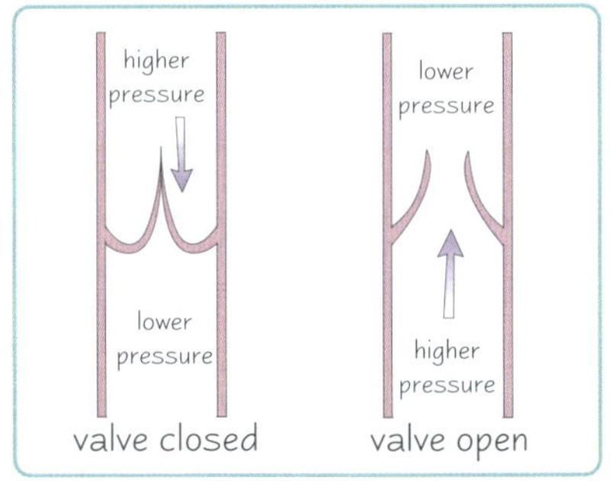

The **valves** only open one way — whether they open or close depends on the relative pressure of the heart chambers. If there's higher pressure behind a valve, it opens easily, but if pressure is higher above the valve then it closes.

Each bit of the heart is adapted to do its job effectively.

1) The **left ventricle** of the heart has thicker, more muscular walls than the **right ventricle**, because it needs to contract powerfully to pump blood all the way round the body. The right side only needs to get blood to the lungs, which are nearby.
2) The **ventricles** have thicker walls than the **atria**, because they have to push blood out of the heart whereas the atria just need to push blood a short distance into the ventricles.
3) The **tricuspid** and **bicuspid** valves link the atria to the ventricles and stop blood getting back into the atria when the ventricles contract.
4) The **semilunar valves** stop blood flowing back into the heart after the ventricles contract.
5) The **cords** attach the atrioventricular valves to the ventricles to stop them being forced up into the atria when the ventricles contract.

Circulation, the Heart and the Cardiac Cycle

The **Cardiac Cycle** Pumps Blood Round the Body

The cardiac cycle is an ongoing sequence of **systole** (contraction) and **diastole** (relaxation) of the atria and ventricles that keeps blood continuously circulating round the body. The systole and diastole alter the **volume** of the different heart chambers, which alters **pressure** inside the chambers. This causes **valves** to open and close, which directs the **blood flow** through the system. There are 3 stages:

1. **Ventricular diastole, atrial systole**

   The **ventricles both relax**. The atria then contract, which decreases their volume. The resultant higher pressure in the atria causes the atrioventricular valves to open. This forces blood through the valves into the ventricles.

2. **Ventricular systole, atrial diastole**

   The **atria relax** and the **ventricles then contract**. This means pressure is higher in the ventricles than the atria, which shuts off the atrioventricular valves to prevent backflow. Meanwhile, the high pressure opens the semilunar valves and blood is forced out into the pulmonary artery and aorta.

3. **Ventricular diastole, atrial diastole**

   The **ventricles and the atria both relax**, which increases volume and lowers pressure in the heart chambers. The higher pressure in the pulmonary artery and aorta closes the semilunar valves to prevent backflow. Then the atria fill with blood again due to higher pressure in the vena cava and pulmonary vein and the cycle starts over again.

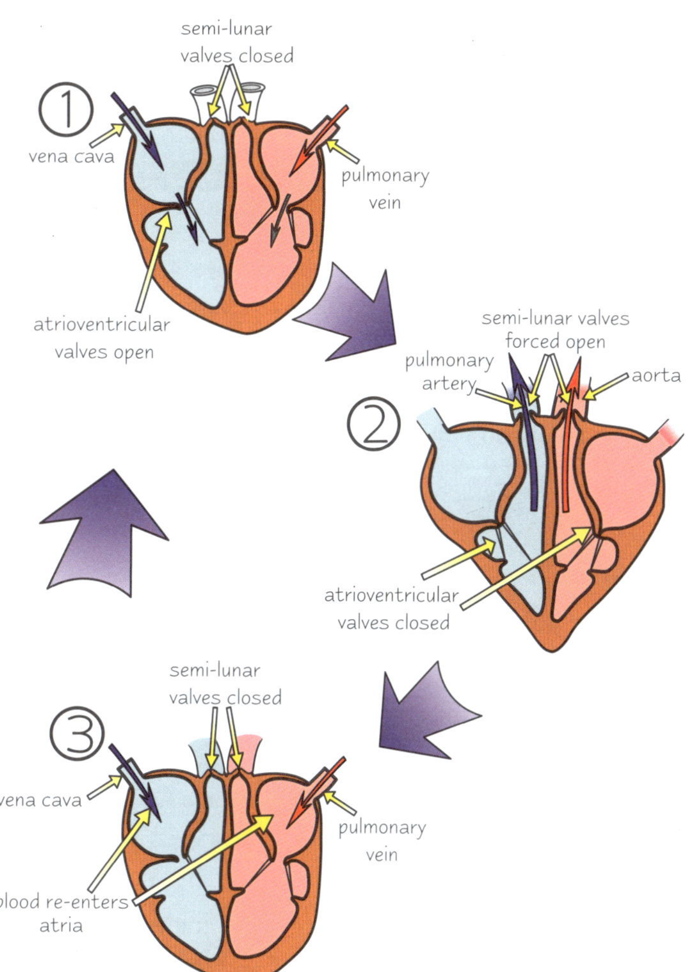

Practice Questions

Q1 What is the difference between the systemic and pulmonary circulatory systems?

Q2 Which side of the heart carries oxygenated blood?

Q3 Why is the left ventricle more muscular than the right ventricle?

Q4 What is the purpose of heart valves?

Exam Questions

Q1 Describe the pressure changes which occur in the heart during systole and diastole. [3 marks]

Q2 Explain how valves stop blood going back the wrong way. [6 marks]

Learn these pages off by heart...

Some of this will be familiar to you from GCSEs — so there's no excuse for not learning it really well. The diagram of the heart can be confusing — it's like looking at a mirror image, so right is left and left is right. (So in fact, when you look in the mirror you don't see what you actually look like — you see a reverse image — weird.)

Blood Vessels

Blood vessels are a welcome relief from that last page, methinks. All you've got to think about here is which tube is thicker than which, and what carries what. It's not hard. In fact, it's a doddle. A page to savour and enjoy.

There are **Five** Main Types of **Blood Vessel**

The five main types of blood vessels are **arteries**, **arterioles**, **capillaries, venules** and **veins**.

1) **Arteries** carry blood **from** the heart **to** the rest of the body. They're thick-walled, muscular and have elastic tissue in the walls to cope with the **high pressure** caused by the heartbeat. All arteries carry **oxygenated** blood except the **pulmonary arteries**, which take deoxygenated blood to the lungs.

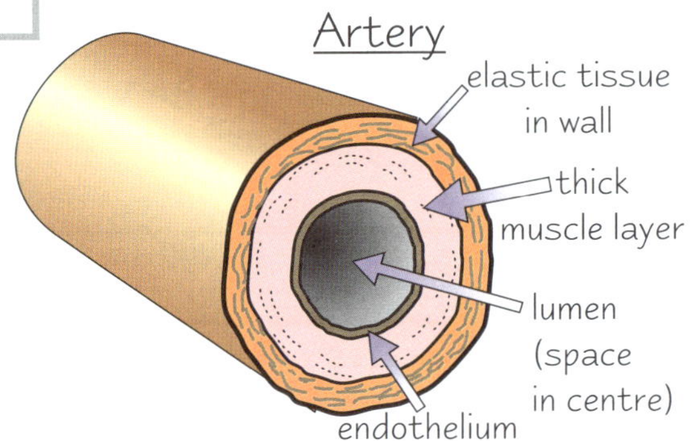

2) Arteries divide into smaller **arterioles** which form a network of vessels throughout the body. Blood is directed to different **areas of demand** in the body by muscles inside the arterioles contracting and restricting the blood flow or relaxing and allowing full blood flow.

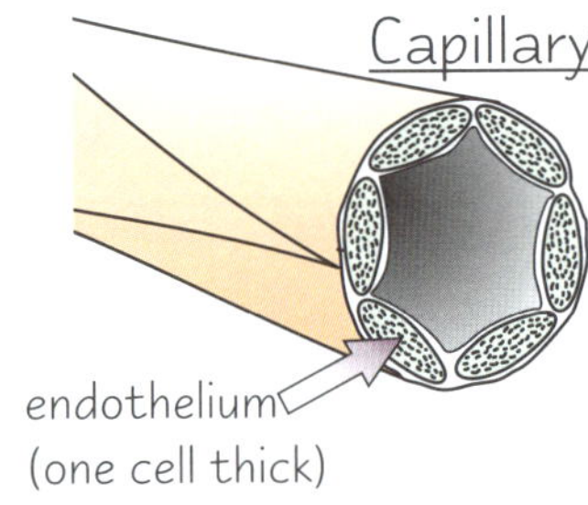

3) Arterioles branch into **capillaries**, which are the **smallest** of the blood vessels. Substances are exchanged between cells and capillaries. Their walls are only **one cell thick** to allow efficient **diffusion** of substances (e.g. glucose and oxygen) to occur near cells. At their other end, capillaries merge into venules (see below). Networks of capillaries in tissue are called **capillary beds**.

4) **Venules** are a bit bigger than capillaries and form a network of vessels which spread away from capillary beds towards veins.

5) **Veins** take blood back **to the heart**. They're **wider** than equivalent arteries, with very little elastic or muscle tissue. Veins contain **valves** to stop the blood flowing backwards. Blood flow through the veins is helped by contraction of the **body muscles** surrounding them. All veins carry **deoxygenated** blood (because oxygen has been used up by body cells), except for the **pulmonary veins**, which carry oxygenated blood to the heart from the lungs.

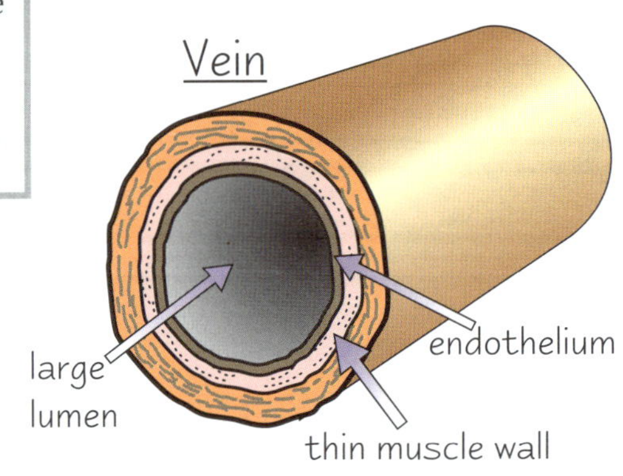

Control of Heartbeat

You don't have to think consciously about making your heart beat — your body does it for you. So you couldn't stop it beating even if for some strange reason you wanted to. Which is nice to know.

Cardiac Muscle Controls the Regular Beating of the Heart

Cardiac muscle is '**myogenic**' — this means that, rather than receiving signals from **nerves**, it contracts and relaxes on its own. This pattern of contractions controls the **regular heartbeat**.

1) The process starts in the **sino-atrial node (SAN)** in the wall of the **right atrium**.
2) The SAN is like a pacemaker — it sets the rhythm of the heartbeat by sending out regular **electrical impulses** to the atrial walls.
3) This causes the right and left atria to contract **at the same time**.
4) A band of non-conducting **collagen tissue** prevents the electrical impulses from passing directly from the atria to the ventricles.
5) Instead, the **atrio-ventricular node (AVN)** picks up the impulses from the SAN. There is a **slight delay** before it reacts, so that the ventricles contract **after** the atria.
6) The AVN generates its own **electrical impulse**. This travels through a group of fibres called the **bundle of His** and then into the finer fibrous tissue in the right and left ventricle walls called **Purkyne tissue**.
7) The impulses mean both ventricles **contract simultaneously**, from the bottom up.

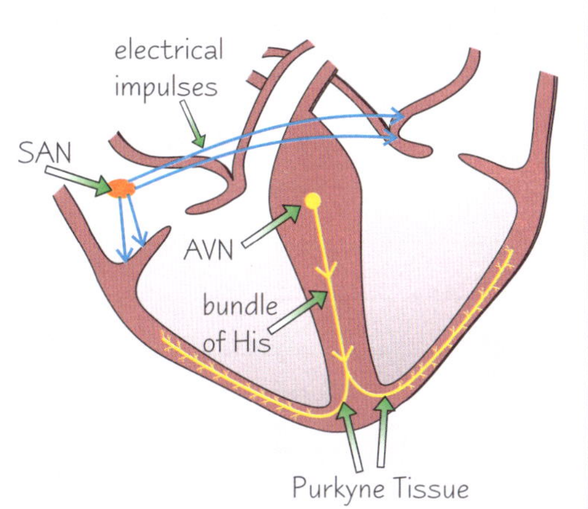

Sometimes the heart rhythm gets out of control. The wave of stimulation to heart muscle becomes chaotic. Different parts of the heart contract and relax at the same time. This is called fibrillation and can be fatal. Luckily, doctors can stop it using defibrillating equipment, which sends an electric shock to the heart, returning it to its proper rhythm.

Practice Questions

Q1 Where do arteries carry blood from and to?

Q2 What is a lumen?

Q3 Which type of blood vessel is only one cell thick?

Q4 Which blood vessels have valves, and why?

Q5 What does myogenic mean?

Q6 What is the difference between SAN and AVN?

Q7 What two types of tissue do electrical impulses from the AVN travel through?

Exam Questions

Q1 Describe the structure of arteries and veins. [6 marks]

Q2 Describe how the heartbeat is stimulated. [6 marks]

What happened to the Bundle of Hers? — she lost her nerve & had to go…

Did you know that your heart can continue beating for a while after you're 'clinically dead' — spooky stuff. The reason is that it doesn't need any input from the brain and nervous system to operate. But, although the brain doesn't control heartbeat it does control the ***rate*** *of the heartbeat.*

Blood and Tissue Fluid

For some people the word 'blood' is enough to make them cringe. Strange really, seeing as it's the substance that keeps us all alive. So sorry to all you blood-phobics out there, because these pages are all about the stuff.

Blood Contains **Blood Cells**, **Platelets** and **Plasma**

Blood is a **specialised tissue** that's composed of 45% **blood cells** and **platelets** suspended in a liquid called **plasma** (55%). Plasma's mainly **water**, with various nutrients and gases dissolved in it. The main role of blood is **transporting substances** around the body **dissolved** in the **plasma**. Substances move into and out of the plasma through **blood capillaries** at exchange surfaces.

Also, antibodies are secreted directly into plasma by white blood cells.

Substance	Exchange surface where substance enters blood through capillaries	Exchange surface where substance leaves blood through capillaries
nutrients from digestion (glucose, amino acids, fatty acids, mineral ions)	epithelium of the villi in the small intestine	body tissues
hormones	glands	target organs
oxygen	alveoli in the lungs	body tissues
carbon dioxide	body tissues	alveoli in the lungs
urea	liver cells	kidney cells

Blood Cells are **Adapted** to Specific Jobs

1) **Red blood cells** (**erythrocytes**) are responsible for absorbing **oxygen** and transporting it round the body. They're made in the **bone marrow** and are very small.

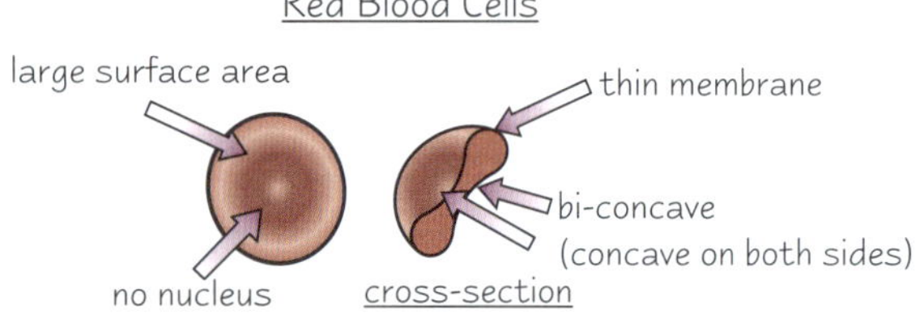

1) They have **no organelles** (including no nucleus) to leave more room for **haemoglobin**, which carries the oxygen.
2) They have a large surface area due to their **bi-concave disc** shape. This allows O_2 to diffuse quickly into and out of the cell.
3) They have an **elastic membrane**, which allows them to change shape to squeeze through the small blood capillaries, then spring back into normal shape when they re-enter veins.

2) **White blood cells** (**leucocytes**) are larger than red blood cells but there are fewer of them in blood. They are responsible for fighting disease.

There are two main types of white blood cell:

White Blood Cell	Diagram	Function and Structure
phagocytes (neutrophils and macrophages)		• engulf pathogens and micro-organisms and digest them (phagocytosis) • contain many lysosomes to aid digestion • elongated nucleus and flowing cytoplasm enables them to squeeze through gaps between cells to move to the site of infection in body tissues
lymphocytes		• produce antibodies to prevent disease • large nucleus

In addition, white blood cells can be divided into two general types — those with a **granular cytoplasm** and those without.

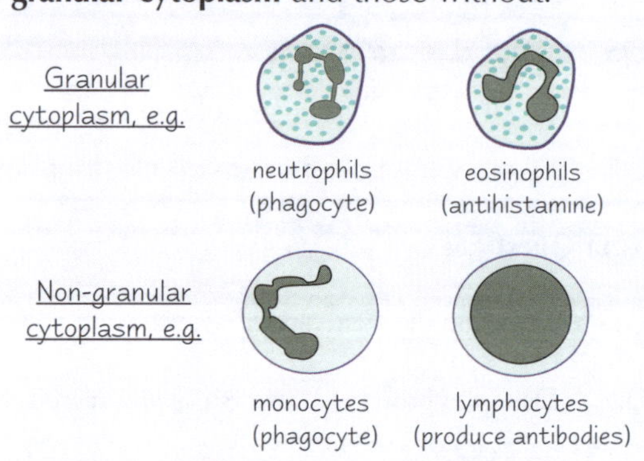

3) **Platelets** are tiny bits of **cytoplasm** held within a cell membrane. They have no nucleus. They release an enzyme that produces **fibrin fibres** to form a **blood clot** when blood vessels are cut open — this is what forms **scabs**.

Blood and Tissue Fluid

Tissue Fluid is Formed from Blood Plasma

Tissue fluid surrounds the cells in tissues — providing them with the conditions they need to function. Tissue fluid is made from substances which leave the plasma from the blood capillaries. Substances move out of blood capillaries by **pressure filtration**:

1) At the **arteriole end** of the capillary bed, pressure inside the capillaries is **greater** than pressure in the tissue fluid. This difference in pressure forces fluid to **leave** the **capillaries** and enter tissue space.
2) As fluid leaves, pressure reduces in the capillaries — so the pressure's much lower at the **venule end** of the capillary bed.
3) Due to the fluid loss, the **water potential** at the **venule end** of the capillaries is **lower** than the water potential in the **tissue fluid** — so some **water re-enters** the capillaries from the tissue fluid at the venule end, by **osmosis**.

Unlike blood, tissue fluid **doesn't** contain **red blood cells** or **big proteins**, because they are **too large** to be pushed out through the capillary walls. It does contain smaller molecules, e.g. oxygen, glucose and mineral ions. Tissue fluid helps cells to get the oxygen and glucose they need, and to get rid of the CO_2 and waste they don't need.

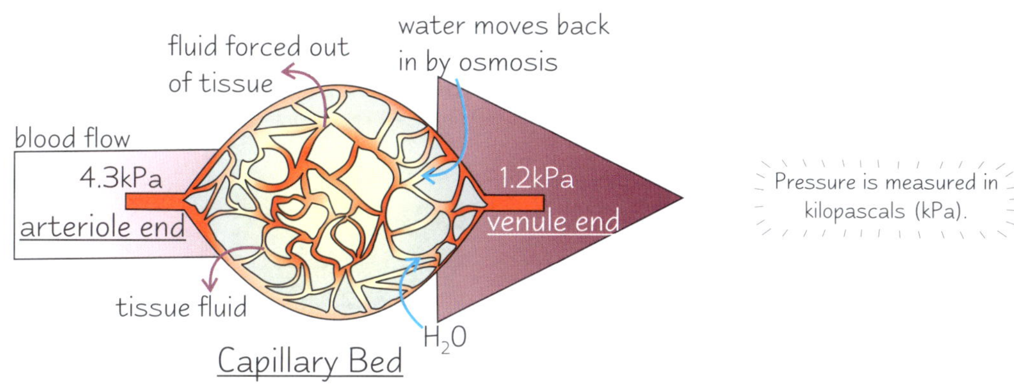

Practice Questions

Q1 What does blood contain?

Q2 Why don't red blood cells have any organelles?

Q3 What do white blood cells do?

Q4 What are the two main types of white blood cell?

Q5 Name a white blood cell with a non-granular cytoplasm.

Q6 What is pressure filtration?

Exam Questions

Q1 How does tissue fluid move into and out of blood capillaries? [5 marks]

Q2 Explain how the structure of phagocytes helps them to fight disease in the body. [3 marks]

That's the end of this bloody topic — and you can't get me for swearing...

It's time to conquer any fears of blood and start appreciating it for the amazing tissue it is. Learn its functions, and which bits do each function — plasma does transport, red blood cells do oxygen and white blood cells do disease-fighting. If I were a blood cell I'd be a great leucocyte warrior, instilling fear into the nuclei of pathogens everywhere. Aaanyway.

Haemoglobin and Oxygen Transport

Aaagh, complicated topic alert. Don't worry though, because your poor, over-worked brain cells will recover from the brain-strain of these pages thanks to oxyhaemoglobin. So the least you can do is learn how it works.

Oxygen** is Carried Round the Body as **Oxyhaemoglobin

Oxygen is carried round the body by **haemoglobin** (Hb), in red blood cells. When oxygen joins to it, it becomes **oxyhaemoglobin**. This is a **reversible reaction** — when oxygen leaves oxyhaemoglobin (**dissociates** from it), it turns back to haemoglobin.

$$\mathbf{Hb} + \mathbf{4O_2} \rightleftharpoons \mathbf{HbO_8}$$

Haemoglobin + oxygen ⇌ oxyhaemoglobin

1) **Haemoglobin** is a large, **globular protein** molecule made up of four polypeptide chains (see p.8).
2) Each chain has a **haem group** which contains **iron** and gives haemoglobin its **red** colour.
3) Haemoglobin has a **high affinity for oxygen** — each molecule carries **four oxygen molecules**.

'Affinity' for oxygen means willingness to combine with oxygen.

Partial Pressure** Measures **Concentration** of **Gases

The **partial pressure of oxygen** (***p*O_2**) is a measure of **oxygen concentration**.
The **greater** the concentration of dissolved oxygen in cells, the **higher** the partial pressure.
Similarly, the **partial pressure of carbon dioxide** (***p*CO_2**) is a measure of the concentration of carbon dioxide in a cell.

Oxygen **loads onto** haemoglobin to form oxyhaemoglobin where there's a **high *p*O_2**.
Oxyhaemoglobin **unloads** its oxygen where there has been a **decrease in *p*O_2**.

1) Oxygen enters blood capillaries at the **alveoli** in the **lungs**. Alveoli cells have a **high *p*O_2** so oxygen **loads onto** haemoglobin to form oxyhaemoglobin.
2) When our **cells respire**, they use up oxygen. This **lowers *p*O_2**, so red blood cells deliver oxyhaemoglobin to respiring tissues, where it unloads its oxygen.
3) The haemoglobin then returns to the lungs to pick up more oxygen.

Athletes train at **high altitude** because there's **low *p*O_2** in the air, which makes the body start producing **more red blood cells**. This means extra oxygen can be carried, so muscles respire more efficiently.

The extra red blood cells stay in the blood for a couple of weeks after returning to normal altitudes, giving the athlete an advantage. Sneaky.

***Dissociation Curves** Show How **Affinity for Oxygen** Varies*

Dissociation curves show how the willingness of haemoglobin to combine with oxygen varies, depending on partial pressure of oxygen (*p*O_2).

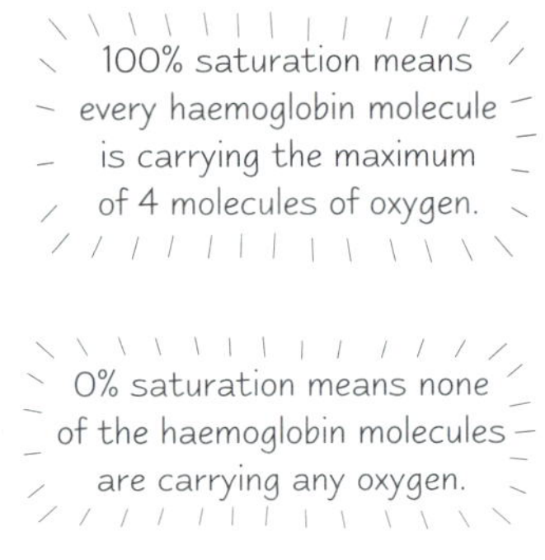

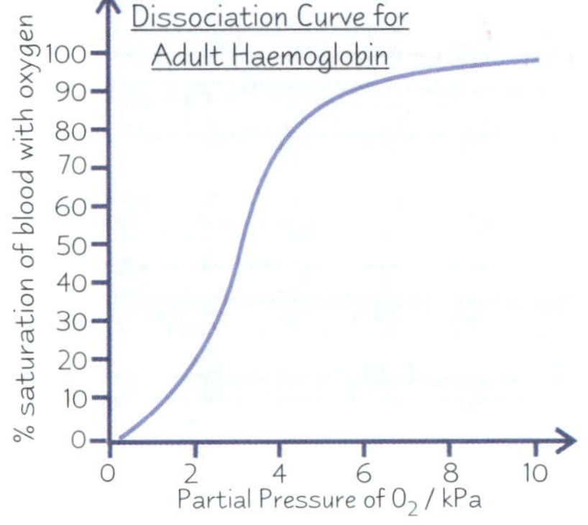

Where ***p*O_2 is high** (e.g. in the lungs), haemoglobin has a **high affinity** for oxygen (i.e. it will **readily combine** with oxygen), so it has a **high saturation** of oxygen.

Where ***p*O_2 is low** (e.g. in respiring tissues), haemoglobin has a **low affinity** for oxygen, which means it **releases oxygen** rather than combines with it. That's why it has a **low saturation** of oxygen.

The graph is '**S-shaped**' because when haemoglobin (Hb) combines with the **first O_2 molecule**, it **alters the shape** of the Hb molecule in a way that makes it **easier** for other molecules to join too. But as the haemoglobin starts to become fully saturated, it becomes harder for more oxygen to join. As a result, the curve has a **steep** bit in the middle where it's really easy for oxygen molecules to join, and **shallow** bits at each end where it's harder for oxygen molecules to join.
When the curve is steep, a small change in *p*O_2 causes a big change in the amount of oxygen carried by the haemoglobin.

Haemoglobin and Oxygen Transport

Different **Blood Pigments** *have Different* **Affinities** *for* **Oxygen**

1) **Foetal haemoglobin**:

In the **womb**, the foetus gets oxygen from its **mother's blood** across the placenta. This can only happen if its blood is **more likely** to absorb oxygen than its mother's blood. Because of this, **foetal haemoglobin** has a **higher affinity for oxygen** than adult haemoglobin. This means that foetal haemoglobin **always** has a slightly **higher saturation of oxygen** than adult haemoglobin, as you can see on the **graph**.

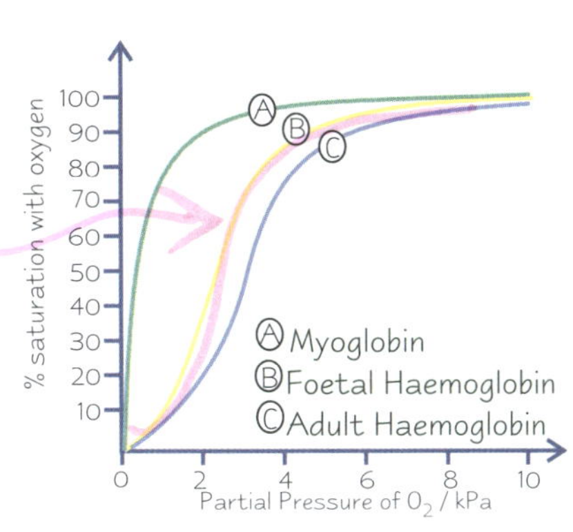

2) **Myoglobin**:

Myoglobin is a **blood pigment** that stores emergency supplies of oxygen in **muscle tissue**. It only releases this oxygen when pO_2 drops to **very low levels**, for example when muscles are using up oxygen faster than haemoglobin can replace it. You can see on the **graph** that its saturation of oxygen remains high, except when pO_2 is very low.

Carbon Dioxide *Levels Affect* **Oxygen Unloading**

To complicate matters, haemoglobin gives up its oxygen **more readily** at **higher partial pressures of carbon dioxide** (pCO_2). It's a cunning way of getting more oxygen to cells during activity. When cells respire they produce carbon dioxide, which raises pCO_2, increasing the rate of oxygen unloading. The reason for this is linked to how CO_2 affects blood pH.

1) CO_2 from respiring tissues diffuses into red blood cells and is converted to **carbonic acid**.
2) The carbonic acid **dissociates** to give **hydrogen ions** and **hydrogencarbonate ions**.
3) If left alone, the hydrogen ions would increase the cell's acidity. To prevent this, oxyhaemoglobin **unloads** its oxygen so that haemoglobin can take up the hydrogen ions.
4) The **hydrogencarbonate ions** diffuse out of the red blood cells and are **transported in the plasma**.
5) When the blood reaches the **lungs** the low concentration of CO_2 causes the hydrogencarbonate and hydrogen ions to **recombine into** CO_2.
6) The CO_2 then diffuses into the **alveoli** and is breathed out.

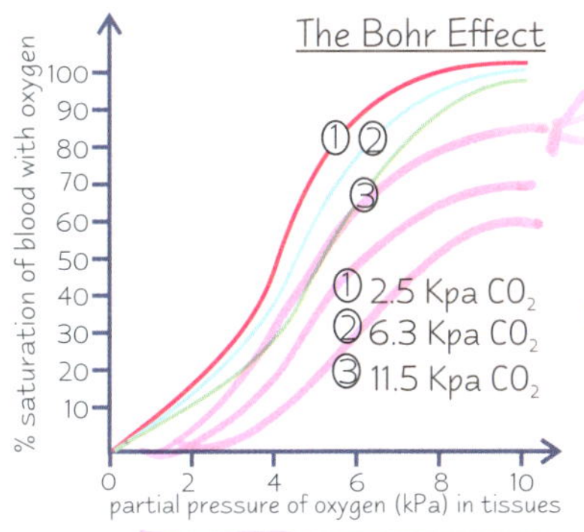

When carbon dioxide levels increase, the dissociation curve 'shifts' down, showing that more oxygen is released from the blood (because the lower the saturation of O_2 in blood, the more O_2 is being released). This is called the Bohr effect.

Practice Questions

Q1 How is HbO_8 formed?

Q2 What is pO_2?

Q3 What are the differences between myoglobin and haemoglobin?

Q4 What is carbon dioxide converted to in red blood cells?

Exam Questions

Q1 Explain why foetal haemoglobin is different from adult haemoglobin. [3 marks]

Q2 Explain the importance of the Bohr effect during a marathon race. [4 marks]

The Bore effect — it's happening right now...

Dissociation graphs can be a bit confusing — but basically, when tissues contain lots of oxygen (i.e. pO_2 is high), haemoglobin readily combines with the oxygen, so blood has a high saturation of oxygen (and vice versa when pO_2 is low). Simple. Also, make sure you get the lingo right, like 'partial pressure' and 'affinity' — hey, I'm hip, I'm groovy.

Environmental Adaptations

Organisms have adaptations, which help them survive in their particular environments. Like, my tiny brain means I'm well adapted to write stupid introductions and top tips for you. Which I'm sure you're delighted about.

Organisms Can be Adapted to their Environment in Three Ways

Organisms are adapted to make the most of their **surroundings** in three different ways:

1) **Structural adaptations** — the **size and shape** of the organism, the structure of any **appendages** (e.g. limbs, tail, ears, etc.), whether it has fur, feathers, scales, etc. e.g. fish have fins and a streamlined body to help them swim in water.

2) **Physiological adaptations** — how the **cells**, **organs** and **organ systems** are made up, e.g. plants that live in places of high salinity have special cells of even higher salt concentration than their surroundings to allow water uptake by osmosis.

3) **Behavioural adaptations** — how the organism acts, e.g. some animals hibernate in winter to avoid the low temperatures and lack of food.

Darwin's **theory of natural selection** explains how organisms become adapted to their environment.

Gene mutations arise by chance. Some of these mutations code for **characteristics** that improve an organism's chances of survival in a particular environment.

↓

The organisms with the favourable characteristic are likely to **survive longer**, reproduce more, and pass their **genes** on to more offspring.

↓

This means that more of the next generation **inherit** the favourable characteristic.

↓

Over time, the whole species may gradually change its characteristics by this process of **natural selection**. Sometimes **new species** can arise.

Adaptations of Organisms Relate to Their Environment

Many plants have features to **prevent water loss** by transpiration. They're called **xeromorphic** adaptations. You find them in **xerophytes** — plants that live in warm, dry or windy habitats, which speed up **transpiration rate**. Examples include:

1) Thicker **waxy cuticles** on leaves and stems.
2) Leaves are reduced to **spines**. This decreases the number of stomata and lowers the surface area from which water can be lost, e.g. **cacti** have spines for leaves.
3) Leaves can **roll up** in especially dry conditions — e.g. **marram grass**. This reduces the surface area for losing water and traps moist air, slowing down transpiration.
4) Layer of '**hairs**' on the epidermis traps moist air round the stomata, to reduce the diffusion gradient.
5) **Swollen stem** to store water, and **roots** spread over a **wide area** just below the soil surface, to make the most of any rain.

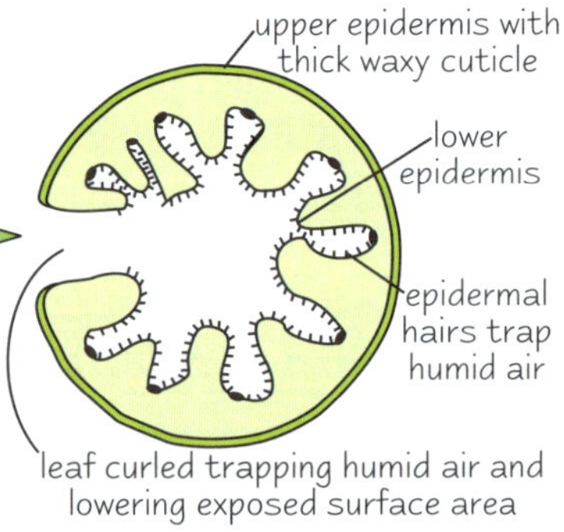

Other plants live in areas where there's **lots of water**. These are called **hydrophytes**. Hydrophytes that are **fully submerged** in fresh water have adaptations like:

1) No cuticle or stomata.
2) Reduced **supporting tissue** to help them move with the water currents.
3) **Air bladders** to help them float nearer the surface where there's higher light intensity.

Animals also have external adaptations to help them survive. For example, **moles** live underground:

1) They have tiny eyes, because vision isn't so important underground.
2) Their forelimbs are designed to shovel the earth.

Other examples are otters, seals and dolphins, which have streamlined bodies to help them swim efficiently.

Environmental Adaptations

Invertebrates are *Adapted* to the *Oxygen Level* in their *Environment*

Oxygen is essential to organisms that carry out **aerobic respiration**, and oxygen levels differ between habitats. **Invertebrates** have **physiological** and **structural** adaptations for gas exchange that reflect the amount of oxygen in their environment.

1. In **terrestrial habitats** there's **direct access to air**. Some invertebrates are adapted by having **breathing tubes** or **lungs** to obtain oxygen. **Insects** use a **tracheal system** — surface **air pores** lead to a network of **tracheal tubes**, kept open by a hardened lining. These in turn lead to a system of very fine tubes (**tracheoles**) that branch deep into all the tissues, and are unlined to allow gas exchange. Oxygen **diffuses directly** into the tissues with no need for any circulatory system.

2. There's generally **less oxygen** in **aquatic habitats**, but there's more in **moving water** (e.g. a fast-flowing river) than in **stagnant** water (e.g. the bottom of a pond). The oxygen content can also **vary** a lot, depending on the **temperature** (there's **less** at **lower** temperatures). Some aquatic invertebrates obtain oxygen by **diffusion** through their whole body surface. This needs a large **surface area to volume ratio** (see p.34). Others have different adaptations:

INVERTEBRATE	ADAPTATIONS FOR GETTING OXYGEN
Amoeba	**Single-celled** organism, so surface area to volume ratio is high.
Flatworms (*platyhelminthes*), e.g. the tapeworm	**Flattened body shape** increases surface area to volume ratio.
Annelids, e.g. the leech	As well as receiving oxygen through diffusion, they have a simple **circulatory system** to get it to all their internal cells. Their blood contains **haemoglobin** to increase the **oxygen carrying capacity**. Not all invertebrates have haemoglobin in their blood.
The sludge worm (*Tubifex*), a type of aquatic **oligochaete**	They live at the bottom of rivers in **low oxygen concentrations**. • They bury their heads in the mud and leave their tails sticking up. • They then absorb oxygen through the surface of their **tail ends**, which **wave about** to move **oxygenated water** over them. • The blood **capillaries** in their tail ends are close to the surface, to make diffusion easier. • Their blood contains **haemoglobin**.
Lugworms	They have **external gills**, which allow more blood to flow close to the water so that gases can be exchanged more easily.

Practice Questions

Q1 What is meant by
a) structural adaptations?
b) physiological adaptations?
c) behavioural adaptations?

Q2 Name three external xeromorphic adaptations.

Q3 What is a hydrophyte?

Exam Questions

Q1 Name one behavioural, one structural and one physiological adaptation to low oxygen concentration shown by a tubifex worm. [3 marks]

Q2 The oxygen concentration was measured on the same day in three habitats — a fast-flowing river, a pond and a terrestrial environment. The three concentrations were 10 $cm^3 l^{-1}$, 210 $cm^3 l^{-1}$, and 5 $cm^3 l^{-1}$. Match the most likely habitat to each oxygen concentration and explain your answer. [5 marks]

Xeromorphic — an exciting word for a boring subject...

Actually that's unfair. It's taken millions of years for plants to evolve those adaptations, and here I am slagging them off. When I've managed to develop a thicker waxy cuticle on my leaves and stems, then I can comment, and not before. Oh, and learn this page. It may not be thrilling — but if you know it, it could earn you vital marks.

Meiosis and Sexual Reproduction

*More cell division — lovely jubbly. Meiosis is the cell division used in sexual reproduction. It consists of two divisions, not one. The **second division** is exactly the same as mitosis (see p.32-33), which is handy.*

DNA From One Generation is Passed to the Next by Gametes

1) **Gametes** are the **sperm** cells in males and the **ova** (egg cells) in females. They join together at **fertilisation** to form a **zygote**, which divides and develops into a **new organism**.
2) Normal **body cells** have the **diploid number** (**2n**) of chromosomes — meaning each cell contains **two** of each chromosome, one from the mum and one from the dad.
3) **Gametes** have a **haploid** (**n**) number of chromosomes — there's only one copy of each chromosome.
4) At **fertilisation**, a **haploid sperm** fuses with a **haploid egg**, making a cell with the normal diploid number of chromosomes. The new organism gets half its chromosomes from the father (the sperm) and half from the mother (the egg).

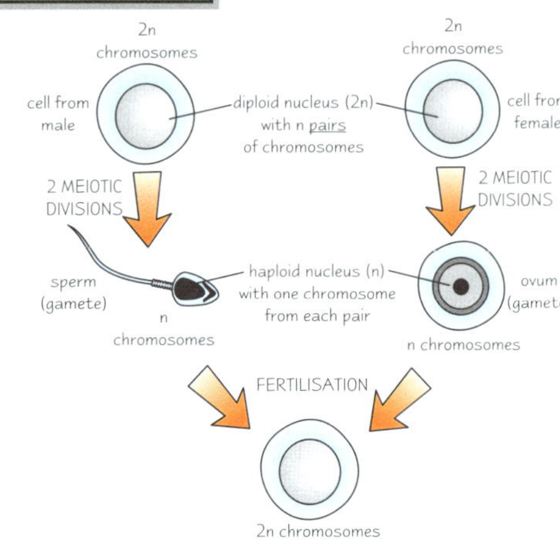

Meiosis Halves the Chromosome Number

1) **Meiosis** is a type of cell division essential for **sexual reproduction**. Cells that are **diploid** to start with divide into **haploid** cells. Without meiosis, you'd get **double** the number of chromosomes in each generation, when the gametes fused.
2) Meiosis happens in the **reproductive organs**. In humans it's in the **testes** for males and the **ovaries** for females. In plants, it's in the **anthers** and **ovules**.
3) Unlike mitosis, there are **two divisions**. This **halves** the chromosome number.

The two divisions in meiosis are called **meiosis I** and **II**.
Each division has 4 stages, just like mitosis — **prophase**, **metaphase**, **anaphase** and **telophase**.

1) In **Meiosis I** the **homologous pairs** of **chromosomes** are separated, which **halves** the number of chromosomes in the daughter cells.
2) **Meiosis II** is like mitosis — it separates the **pairs of chromatids** that make up each chromosome.
3) Unlike mitosis, which results in two genetically identical diploid cells, meiosis results in **four haploid cells** (gametes) that **are genetically different** from each other.

The **life-cycle** of an organism can be divided into the **diploid phase**, which ends with **meiosis**, and the **haploid phase**, which ends with **fertilisation**. For humans and most other organisms, this means that the diploid phase covers everything from **zygote** to **adult**, and the haploid phase only applies to the **gametes**.

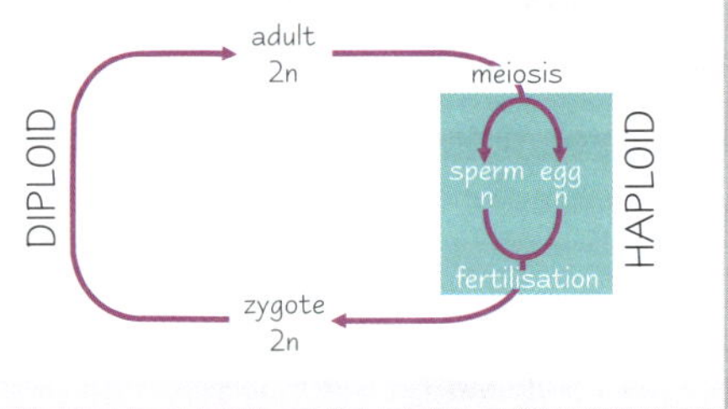

Chromatids Cross-Over During Prophase in Meiosis I

Crossing-over helps make cells from meiosis **non-identical**.

At the start of the prophase in meiosis I (called **prophase I**), **homologous pairs** of chromosomes come together and pair up to form a **bivalent**.
The chromatids twist around each other and fragments of **non-sister chromatids** (chromatids belonging to different chromosomes) swap over. The place where they swap over is called the **chiasma** (plural = **chiasmata**).

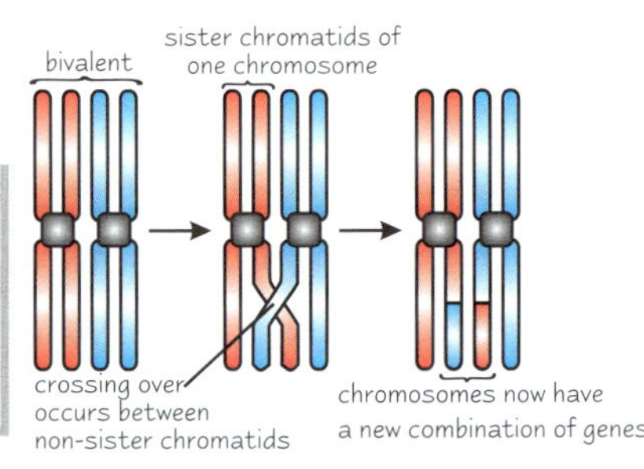

Meiosis and Sexual Reproduction

Meiosis is the Key to **Genetic Variation**

There are two main events during meiosis that lead to **genetic variation**:

1) The four daughter cells formed from meiosis have completely different combinations of **chromosomes**. This is because of **random segregation** of chromosomes in meiosis I, and of the chromatids in meiosis II.

In meiosis I, homologous pairs of chromosomes line up then separate to opposite ends of the cell, and the cell splits in two. The chromosomes can split up any which way:

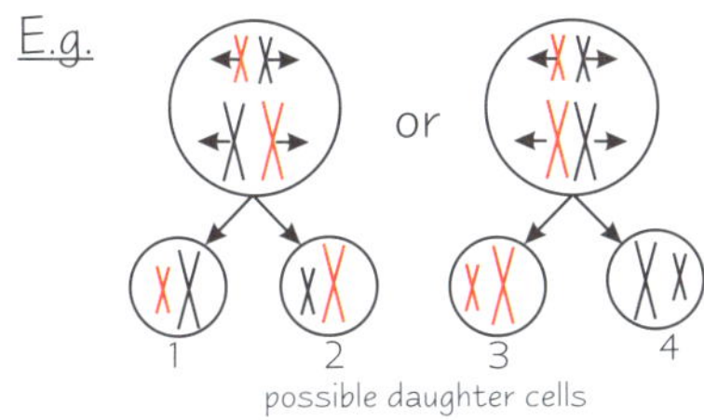

So you get a random combination of chromosomes in the daughter cells. In meiosis II, the chromatids line up and separate to opposite ends of the cell. They can split up any which way too.

2) The **crossing-over** of chromatids in the prophase of meiosis I (**prophase I**) means that each of the **four daughter cells** formed from meiosis contain chromatids with **different genes**:

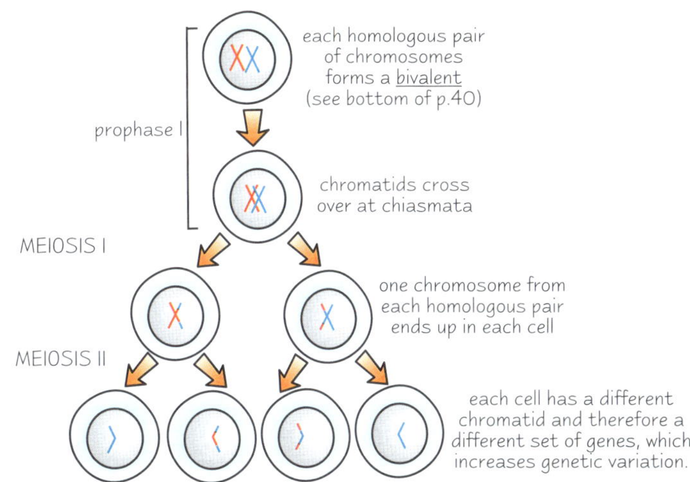

Practice Questions

Q1 Explain what is meant by the terms haploid and diploid.

Q2 When do the haploid and diploid phases of the human life-cycle occur?

Q3 How many divisions are there in meiosis?

Q4 In which organs in a human would meiosis take place?

Exam Question

Q1 The diagram shows stages of meiosis in a human ovary. Each circle represents a cell.

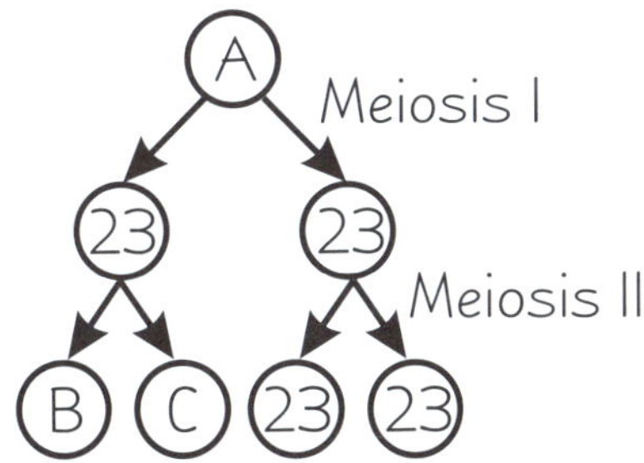

a) How many chromosomes would be found in cells A, B and C? [3 marks]

b) Explain why it's important for gametes to have half the number of chromosomes as normal body cells. [2 marks]

Reproduction isn't as exciting as some people would have you believe...

*For some reason, this stuff can take a while to go in (insert own joke). But that's no excuse to just sit there staring frantically at the page and muttering "I don't get it," over and over again. Use the diagrams to help you understand — they look evil, but they really help. The key thing is to understand what happens to the **number of chromosomes** in meiosis.*

Sexual Reproduction in Plants

Flowers are quite timid about sex. They have to rely on a passing bee or a handy gust of wind to break the ice. But once they get going, there's no keeping that pollen tube inside its grain. Saucy.

Pollination is the Transfer of Pollen Between Flowers

Pollination is transfer of pollen from an **anther** to a **stigma** of flowers of the **same species**. It's NOT the same as fertilisation — don't get these two confused. There are **two types** of pollination:

1) **Self-pollination** — pollen is transferred to the stigma of the **same flower**. This is possible because most flowers are **hermaphrodites** — they have both female **and** male sex organs.
2) **Cross-pollination** — pollen is transferred to the stigma of a **different flower** of the same species.

Cross-pollination leads to greater **genetic variation**, which makes it **preferable** to self-pollination (because variation is needed for **evolution** to take place). Plants have evolved ways to make sure cross-pollination is more likely to happen than self-pollination:

1) **Protandry** — Anthers mature before stigmas, so self-pollination can't occur, e.g. wood-sage.
2) **Protogyny** — Stigmas mature before the anthers, e.g. ribwort plantain.
3) **Dioecious** species — Some species have evolved so that they have either all male parts or all female parts, making self-pollination impossible, e.g. willow.

Most Flowers use either Wind or Insects for Sexual Reproduction

Flowers enable plants to reproduce sexually, but seeing as they can't move they need cunning ways of getting the male gametes (found in pollen) to the female bits of other flowers — no getting jiggy for them. So most use either insects or the wind to do this, and have certain adaptations depending on which method they use.

Insect-pollinated flowers have adaptations to attract insects, such as **brightly coloured petals** and a strong **scent**. Insects land on the flower and pollen grains get stuck to them. They then deposit the pollen onto other flowers they land on.

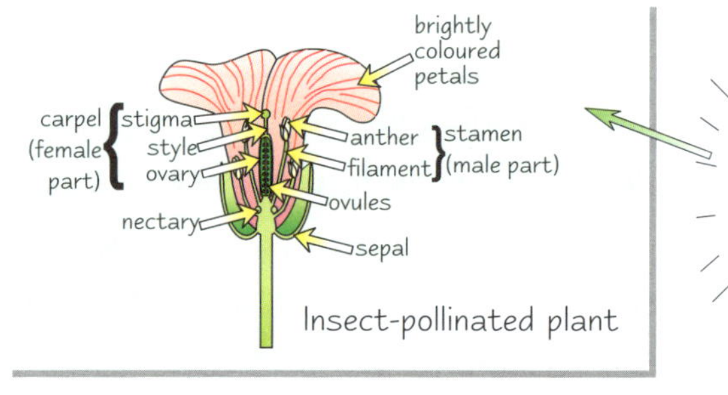

Insect-pollinated plant

The male gametes are found in the pollen grains. They're made in the anther. The female gametes are found in the ovules. They're made in the ovary.

Grasses are **wind-pollinated**. These don't need to attract insects, so they tend to be dull and unscented. Often they don't have petals, leaving the anthers and stigmas **exposed** to the wind. The wind blows pollen grains from the anther of one plant to the stigma of another plant.

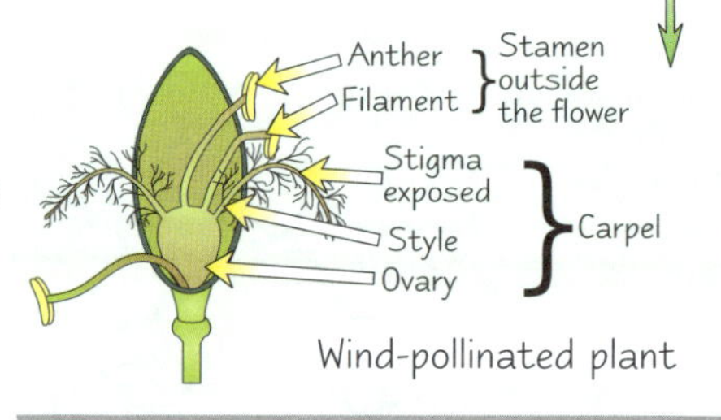

Wind-pollinated plant

The table summarises the characteristic features of a typical wind-pollinated and insect-pollinated flower.

Conventional dating techniques don't work for flowers.

FEATURE	INSECT-POLLINATED	WIND-POLLINATED
Petals	Large, brightly coloured and shiny.	Inconspicuous, dull and small.
Sepals	Present beneath the petals.	Small or absent.
Stamen (male part)	Sturdy and found inside flower.	Flimsy, with large anthers and long filaments hanging out of flower.
Carpel (female part)	Sticky and enclosed within the flower.	Stigmas are long and feathery with a large surface area to trap pollen.
Pollen grains	Large and sticky.	Light and dry.
Scent	Present.	Absent.
Nectar	Produced by nectaries to attract insects.	None produced.

Sexual Reproduction in Plants

Fertilisation is the *Fusion* of Male and Female *Gametes*

1) If a compatible pollen grain lands on the stigma of a flower, the grain absorbs water and splits open.
2) A **pollen tube** grows from it down the **style**. There are **three nuclei** in the pollen tube — one **tube nucleus** at the tube's **tip** and two **male gamete nuclei** behind it. The tube nucleus makes **digesting enzymes**. These are released by the tube so that they digest surrounding cells, making a **way through** for the pollen tube.
3) When the tube reaches the **ovary**, it grows through the **micropyle** (a tiny hole in the ovule wall) and into the **embryo sac** within the **ovule**.
4) In the embryo sac, the **tube nucleus disintegrates** and the tip of the pollen tube **bursts**, releasing the two male gamete nuclei. One male nucleus fuses with the **egg nucleus** to give a **diploid zygote**. This divides by mitosis to become the **embryo** of the seed.
5) The second male nucleus fuses with the **polar nuclei** at the centre of the **embryo sac**. This produces a **triploid nucleus**, which becomes a food store of the mature seed (called the **endosperm**).
6) So a **double fertilisation** has taken place (**two** male nuclei have fused with female nuclei). This only happens in flowering plants.

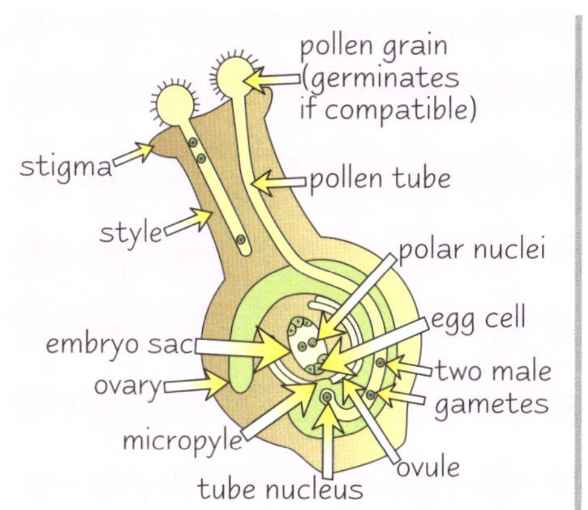

Practice Questions

Q1 What is the difference between self-pollination and cross-pollination?

Q2 Give four features of an insect-pollinated plant.

Q3 Give four features of a wind-pollinated plant.

Q4 How many nuclei are found in a pollen tube?

Exam Questions

Q1 a) The diagram below shows the events leading to fertilisation in a flowering plant. Label parts A-D. [4 marks]

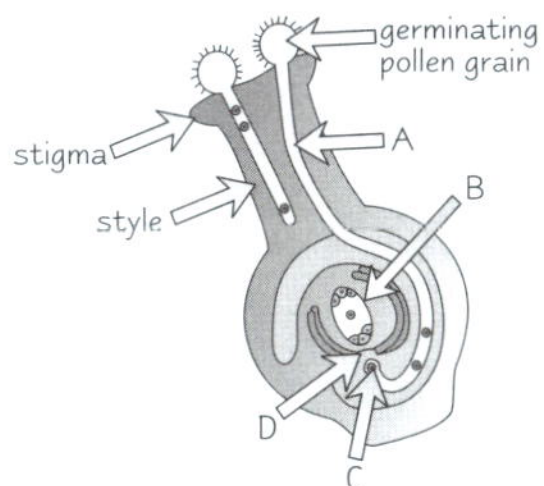

b) Explain why the pollen tube needs to secrete digestive enzymes. [2 marks]

Q2 The diagram shows a picture of a rye grass flower.

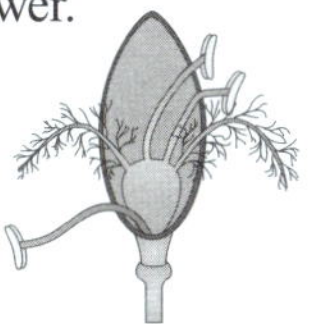

Give two features from the diagram that show it is wind-pollinated. [2 marks]

Sex by wind — and I thought flowers were romantic...

It's easier to remember features if you know why they're there. Flowers pollinated by wind need their anthers sticking out, so the pollen blows away, and their stigmas are feathery to catch pollen blowing past. Insect-pollinated flowers attract insects, which get covered in pollen and carry it to the next flower.

Reproduction in Humans

Yay, it's those diagrams you used to snigger at in Year 7. No time for immature behaviour now though, because you've got to learn about how these systems produce the gametes that make babies. Not a laughing matter. (Tee hee.)

Ova are Produced in the Ovaries of Female Mammals

The female reproductive system produces and develops **ova** (the female gametes), and delivers them to the uterus through the fallopian tubes. It also protects and nourishes the developing **embryo** in the uterus.

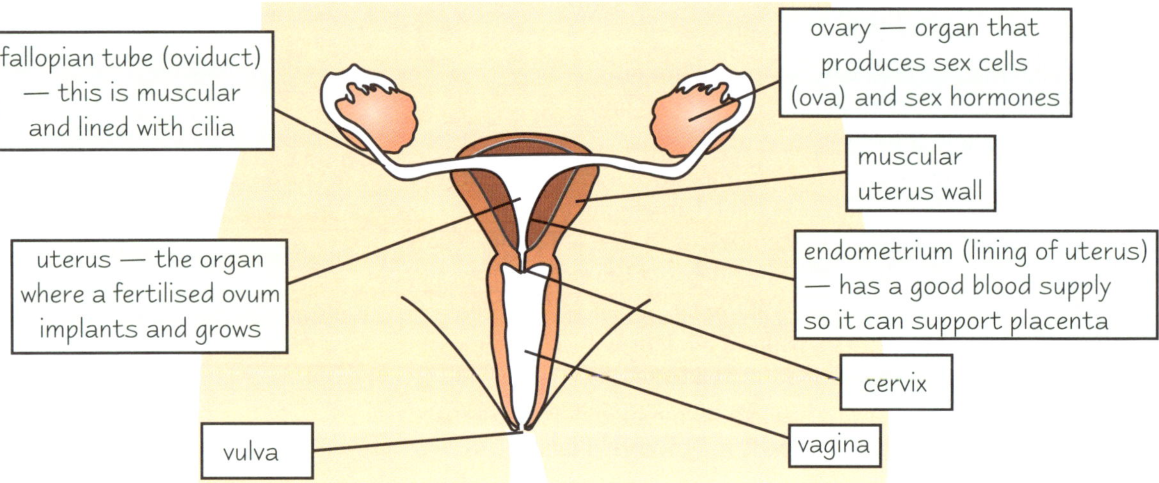

The female reproductive system also produces the female **hormones** that control the menstrual cycle (see p.60 for more on this).

Sperm are Produced in the Testes of Male Mammals

The role of the male reproductive system is to produce the male gametes (sperm) and transfer these into the woman's vagina during sexual intercourse.

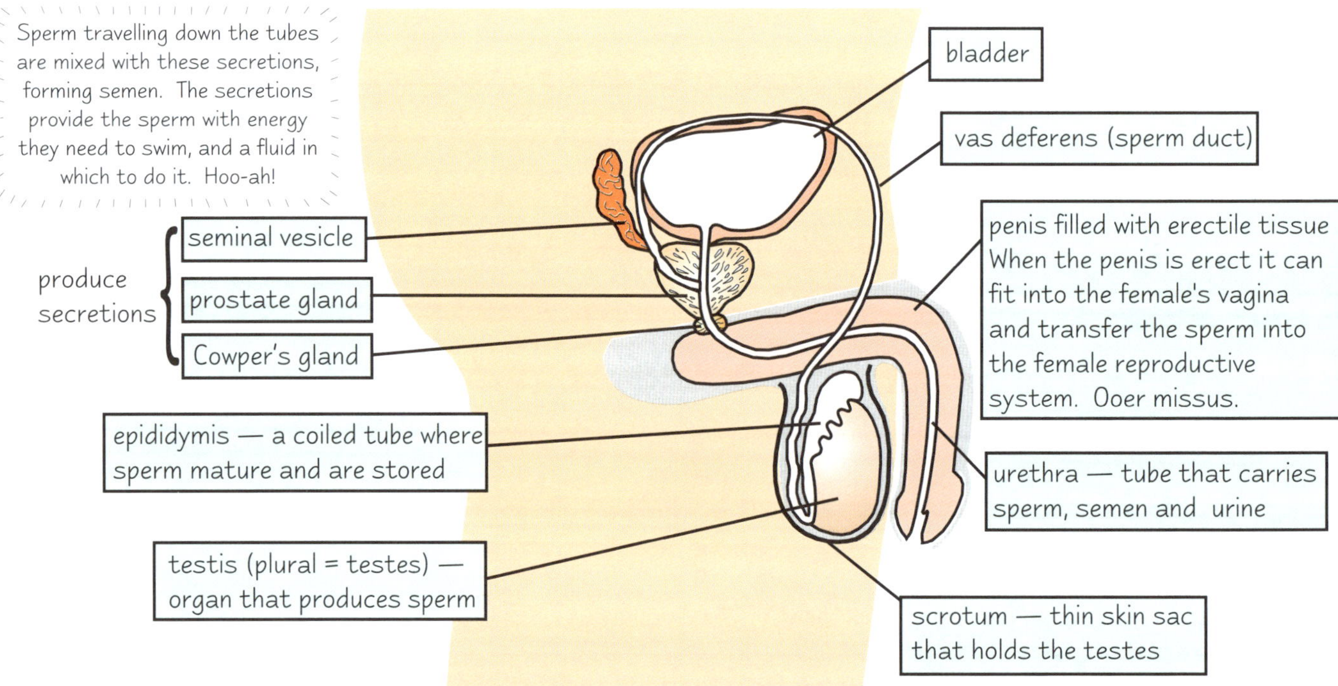

The testes also produce the male hormone, **testosterone**.

Reproduction in Humans

The *Process* that *Makes Ova* is Called *Oogenesis*

Oogenesis begins when the female's still a foetus. Yes, female foetuses begin to make eggs while they're still in the uterus. Odd, I say. Odd.

1) In the foetus, the cells around the outside of the ovaries divide by mitosis to make **oogonia**.
2) The oogonia move towards the middle of the ovary and become **primary oocytes**.
3) These primary oocytes become surrounded by **follicle cells** and are then known as **primary follicles**. When a girl is born, she has millions of these primary follicles, but only some of them develop fully.

Diagram Showing the Development of Ova in the Ovary

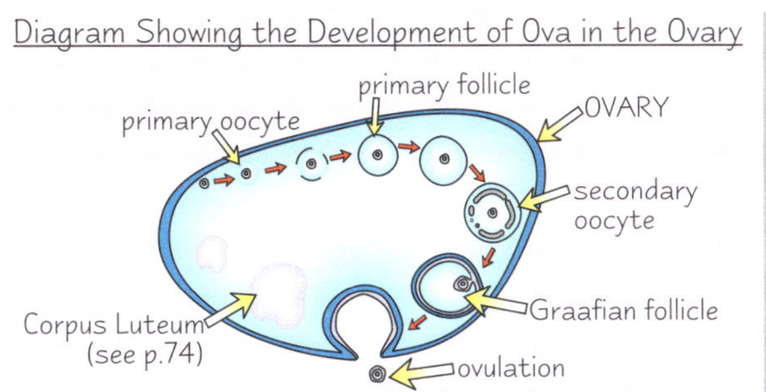

4) At puberty, sex hormones like **FSH** stimulate the primary follicles to divide by meiosis to produce **secondary oocytes** (these are the **ova** that you learn about at GCSE).
5) Each month, one of these secondary oocytes develops inside a **Graafian (mature) follicle**. The follicle travels to the surface of the ovary and bursts, which releases the secondary oocyte into the **fallopian tube**, ready for **fertilisation**. This is called **ovulation**.

The *Process* that *Makes Sperm* is *Spermatogenesis*

Sperm are produced in the walls of **seminiferous tubules** in the **testes** from **puberty onwards**. The process is called **spermatogenesis**.

1) Diploid (2n) cells in the **germinal epithelium** of the tubules divide by **mitosis** many times to give **spermatogonia**.
2) These grow into **primary spermatocytes**.
3) The primary spermatocytes then divide by meiosis to produce haploid (n) **secondary spermatocytes**.
4) The secondary spermatocytes divide once more, to produce **spermatids**. These eventually mature into **sperm**.

Cross-section of a Seminiferous Tubule Wall Showing Spermatogenesis

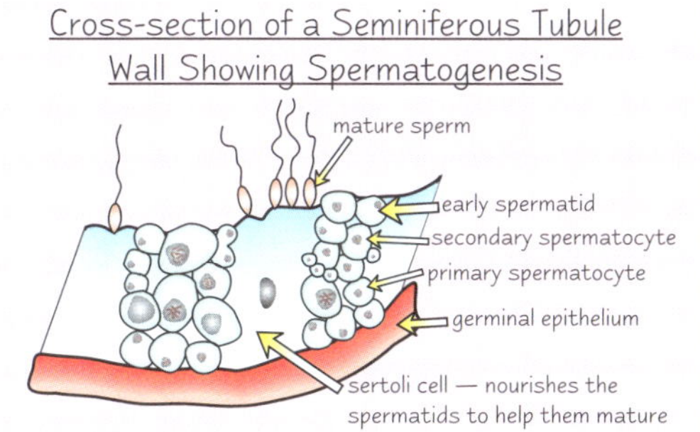

Practice Questions

Q1 Give two functions of the female reproductive system.

Q2 What is the male sex hormone and where is it produced?

Q3 Draw a flowchart to show the stages of oogenesis.

Q4 Where do spermatocytes mature into sperm?

Exam Question

Q1 a) Eggs are produced in the ovaries by a process known as oogenesis. The diagram below shows how a follicle develops in the ovary. Explain what is happening at Y and Z. [2 marks]

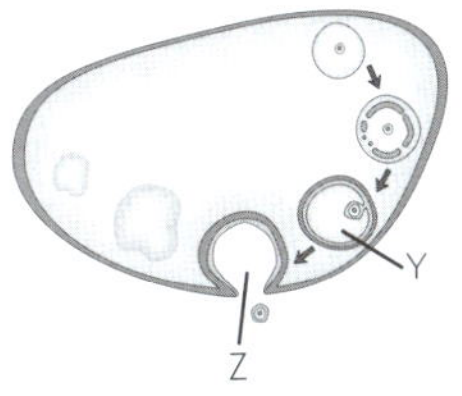

b) Explain how an oocyte reaches the site where fertilisation occurs from where it was produced in the ovary. [3 marks]

Ooooogenesis — putting the 'oooh' into reproduction...

Learn everything on the 'in-yer-face' diagrams — especially all the terms that are new to you, like vas deferens, and epididymis. Don't be put off by oogenesis and spermatogenesis — they're not that complicated really, it's just that the cells go through loads of different stages as they develop, which unfortunately you need to know the names of.

The Menstrual Cycle and Fertilisation

What a lovely page for you to learn. Mmmm... it's all about the menstrual cycle.
Don't worry, boys, there's nothing to be scared of — it's only about random hormones with long names.

*The **Human Menstrual Cycle** is Controlled by **Hormones***

The human menstrual cycle lasts about **28 days**. It involves development of a follicle in the ovary, release of an ovum, and **thickening** of the **uterus lining** so a fertilised ovum can **implant**. If there's no fertilisation, the lining breaks down and exits through the vagina. This is **menstruation**, which marks the end of one cycle and the start of another.

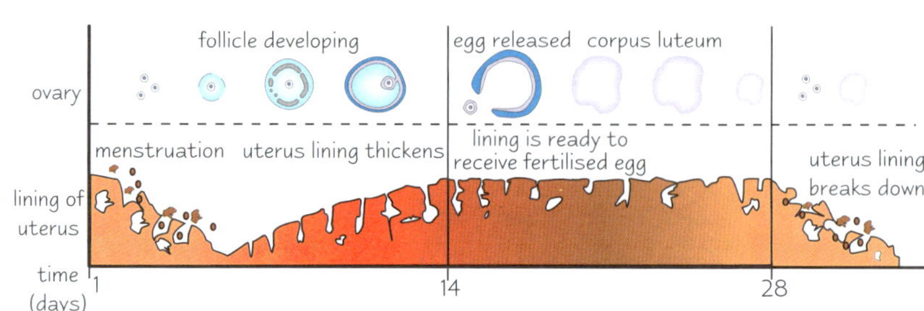

The menstrual cycle is controlled by four hormones — **Follicle-Stimulating Hormone** (FSH), **Luteinising Hormone** (LH), **oestrogen** and **progesterone**. They are either produced in the pituitary gland or in the ovaries:

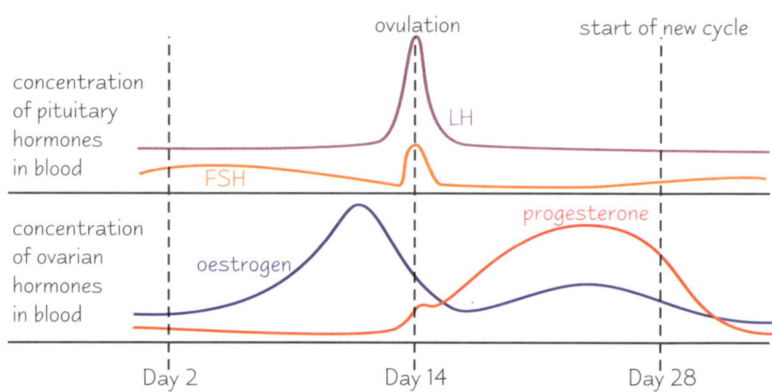

Hormones released by the pituitary —

1) **FSH** is released into the bloodstream at the **start** of the cycle and carried to the ovaries. It stimulates the development of one or more follicles, which stimulates the secretion of oestrogen.
2) **LH** is released into the bloodstream around **day 12**.
It causes release of the ovum (**ovulation**).
When the egg bursts out it leaves its follicle behind.
LH helps the follicle turn into a **corpus luteum**, which is needed later.

Julia couldn't get enough of her cycle.

Hormones released by the ovaries —

1) **Oestrogen** is produced by the **developing follicle**. It causes the **lining** of the uterus to **thicken**. It also **inhibits** release of FSH. This stops any more follicles maturing. The **peak** in oestrogen production **starts** a surge in LH production. The LH **peak** causes ovulation.
2) **Progesterone** is released by the **corpus luteum** after ovulation. Progesterone keeps the lining thick, ready for implantation if fertilisation occurs. It also inhibits release of FSH and LH. If no embryo implants, the corpus luteum dies, so progesterone production stops and FSH inhibition stops. This means the cycle starts again, with development of a new follicle.

The ovum is released around day 14 of the menstrual cycle, and must be fertilised within 24 hours. Best get on with it, then...

The Menstrual Cycle and Fertilisation

A **Sperm** Has to Get to an **Ovum** Before **Fertilisation** Can Happen

1) During **sexual intercourse**, the erect penis fits into the vagina and semen is pushed toward the cervix by **ejaculation**. (Everybody do some deep breathing ... no, not that kind...)
2) By swimming and with help from genital muscular contractions, some sperm will reach the ovum in the fallopian tubes. Fertilisation may then occur.
3) When fertilisation happens, a sperm's **acrosome** releases enzymes which break down the ovum's **zona pellucida**.
4) The sperm passes through, and the cell membranes of the sperm head and ovum merge.
5) The sperm nucleus moves into the cytoplasm, and the two nuclei fuse to form the **zygote**. We're **pregnant**.

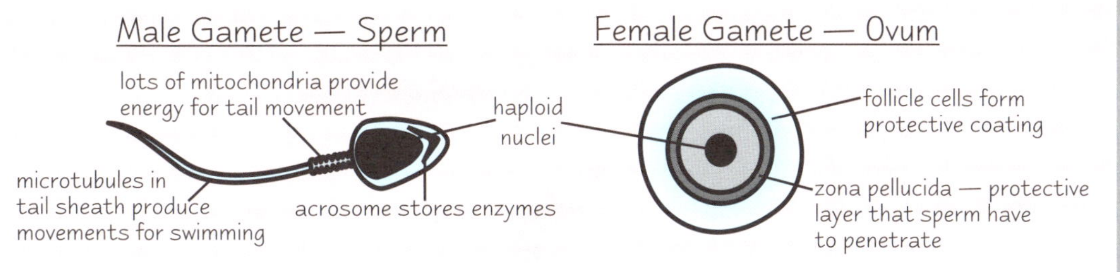

Sperm are small, numerous and mobile. Ova are larger, fewer and can't move independently.

Practice Questions

Q1 Which hormones in the menstrual cycle are released by the pituitary gland? Which are released by the ovaries?

Q2 What does the corpus luteum do?

Q3 Which hormone is released around day 12 of the menstrual cycle?

Q4 Which hormone inhibits the release of FSH?

Q5 Describe how the sperm is adapted to its role in achieving fertilisation.

Q6 What is the zona pellucida?

Q7 Name the part of the sperm that releases enzymes to help with fertilisation.

Exam Question

Q1 Below is a graph showing hormone levels in a woman's blood over 28 days.

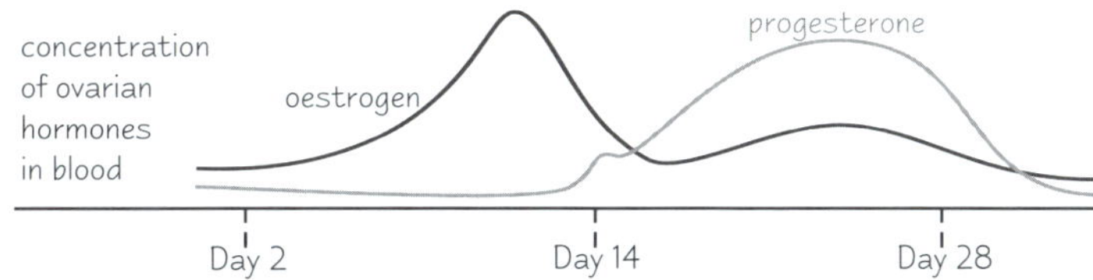

a) Do the hormone levels suggest that this woman might be pregnant? How can you tell? [2 marks]
b) On what day would there be a peak in the concentration of LH? What does this peak cause? [2 marks]
c) What is likely to happen at around day 28? [1 mark]

They used to call it 'the curse' — now I know why...

Lots of hormones. You need to know their names, where they come from, what they do... where their wives go shopping. Make sure you understand how the two diagrams of the menstrual cycle on p.60 link up to each other. It's a lot easier to work out what different hormones do and when if you see it on a diagram rather than written down.

Pregnancy and Birth

There's a lot covered on these pages — implantation of the fertilised ovum, the placenta and pregnancy, birth, lactation... Normally all that stuff takes at least nine months, but I don't recommend you spend that long learning about it.

Implantation *Has to Happen before a* ***Baby*** *Can Develop*

Once an ovum has been fertilised to form a zygote, it needs to reach the uterus and lodge itself in it.

1. **Cilia** and **muscular contractions** of the walls of the fallopian tube push the zygote into the uterus. This takes up to **3 days**.

2. On the way, the zygote divides by **mitosis**, becoming a ball of cells, then a hollow **blastocyst**.

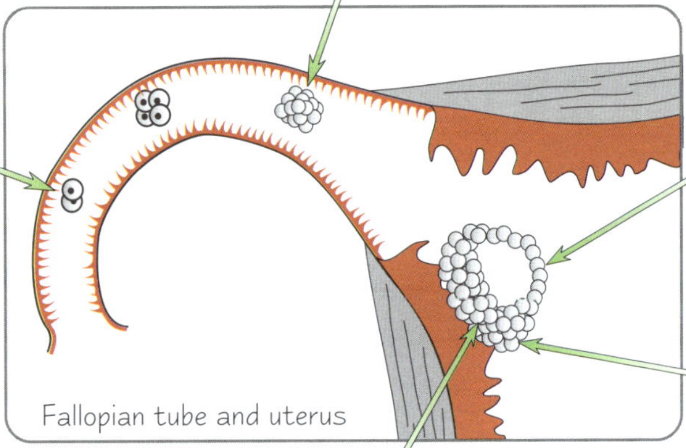

3. The blastocyst **implants** in the **endometrium** (lining of the uterus), which provides nutrition and support at this early stage.

4. The blastocyst secretes **human chorionic gonadotrophin** (HCG), which prevents the **corpus luteum** breaking down, so that **progesterone** is still made and the endometrium stays thick.

5. The blastocyst cells keep dividing so it becomes an **embryo**. The outer layer of blastocyst cells and surrounding endometrial cells divide to form the **placenta**.

Fallopian tube and uterus

The embryo becomes known as the foetus once all the features (e.g. fingers and toes) of a fully-developed baby are recognisable. In humans this happens around the end of the second month of pregnancy.

The ***Placenta*** *is the* ***Link*** *Between* ***Mum*** *and* ***Baby***

The placenta allows the **mother's blood** to flow close to the **foetus's blood**, without them mixing. This stops harmful substances from the mother's blood mixing with the baby's blood, but means that useful substances can **diffuse** between the two blood supplies.

1) The placenta takes over from the corpus luteum, secreting the hormones **progesterone** and **oestrogen**.
2) Oxygen and products of digestion **diffuse** across it from the mother's blood to the foetus's.
3) **Waste products** like carbon dioxide and urea diffuse from the foetus's blood to the mother's.
4) It stops disease-causing **bacteria** getting across from mum to foetus. But some viruses, and **toxins** like alcohol, nicotine and drugs, can still cross.
5) Some **antibodies** can diffuse across, giving the foetus **temporary immunity**.

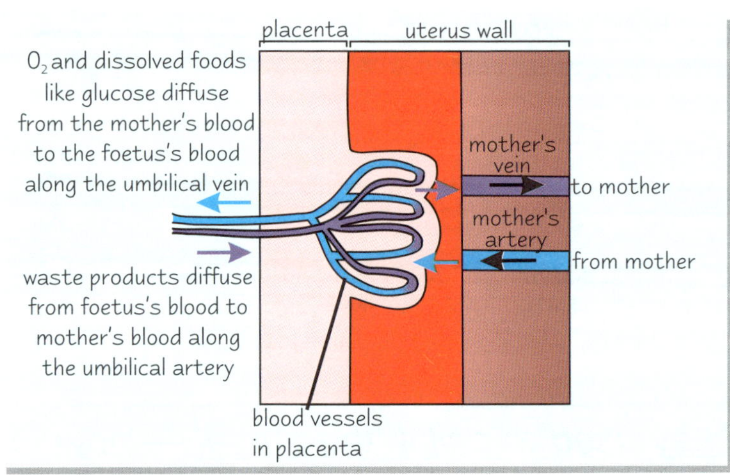

Pregnancy and Birth

Oxytocin and *Prolactin* Control *Birth* and *Lactation*

Gestation — 40 weeks of foetal growth and development in the uterus — is followed by **birth**. **Oxytocin** is a hormone released by the **pituitary gland**. Its main role is to stimulate the smooth muscle in the uterus wall to contract. As more and more is released, the frequency and force of contractions increases. The cervix dilates and the baby is forced out through the vagina, usually head-first.

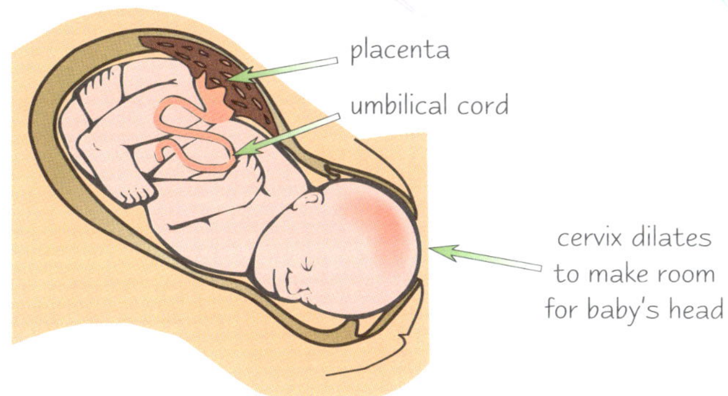

Lactation — At birth, the **pituitary** releases the hormone **prolactin**. This stimulates the breasts to produce milk. The baby's sucking action during breast-feeding stimulates release of **oxytocin** as well as prolactin. Oxytocin causes smooth muscle contractions which expel milk from the breast into the baby's mouth.

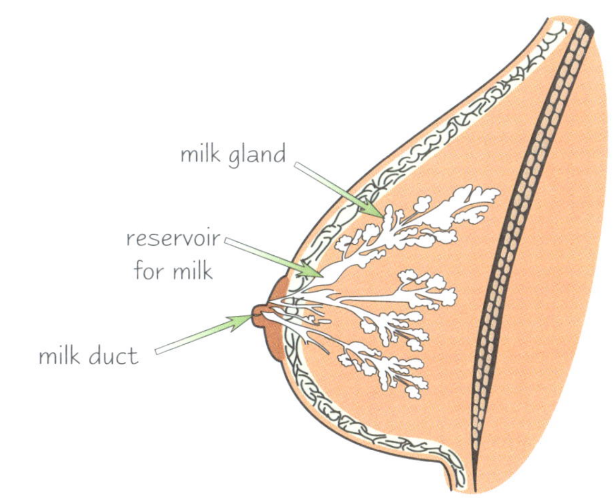

Practice Questions

Q1 Which hormone does the blastocyst secrete?

Q2 What happens when the corpus luteum is prevented from breaking down?

Q3 Give three functions of the placenta.

Q4 Which hormone is responsible for the contractions during labour?

Q5 Which hormone stimulates the breasts to produce milk?

Exam Questions

Q1 Describe the functions of the placenta. [5 marks]

Q2 Describe the functions of the following hormones:

a) Oxytocin [3 marks]

b) Human chorionic gonadotrophin (HCG) [2 marks]

c) Prolactin [1 mark]

Aaaaaarrrrggghhh... Only another 15 hours of labour to go...

Some people like to take their baby's placenta home and eat it. Apparently it tastes quite good — makes nice paté I heard. You'll never catch me eating placenta, but I did dissect one once. My Biology teacher at school brought in his new baby's placenta and let us cut it up. I'd recommend nagging your teacher to have a baby so you can dissect one too. Fantastic.

Modes of Nutrition

Mmmm, food. Different organisms have different ways of feeding themselves — and some of them don't even use cutlery, for shame.

All Organisms are either Autotrophs or Heterotrophs

The first thing you need to know about nutrition is that **all organisms** are either **heterotrophs** or **autotrophs**.

1. **Autotrophs** are **producers**. They make their own complex organic molecules from simple, inorganic ones, using an external energy source. **Photoautotrophs**, like green plants, use the light energy from the sun as their energy source. **Chemoautotrophs**, like some bacteria, use energy from chemical reactions to build up organic molecules.

2. **Heterotrophs** are **consumers**. Heterotrophic organisms get nutrients from complex, organic molecules. These are digested into small, simple molecules, then built up again to make the molecules the organism needs. So whether they eat plants or animals, heterotrophs depend on **autotrophs** to build the molecules in the first place. Parasites, saprophytes (decomposers) and free-living animals are all heterotrophs.

Holozoic Organisms Digest Food Inside Their Bodies

Holozoic nutrition is where food is digested in a tube in the body — the **alimentary canal**. It happens in **five stages**:

Human nutrition is holozoic. Dude.

- **Ingestion** → Food is taken into the body through the mouth.
- **Digestion** → Large, insoluble molecules are converted to small, soluble ones in the gut.
- **Absorption** → Smaller, soluble molecules are taken into the bloodstream from the gut.
- **Assimilation** → Absorbed products of digestion are converted into complex molecules and used by the body.
- **Egestion** → Undigested food is passed out of the body.

Ruminants (animals that eat grass and 'chew the cud', e.g. sheep) and **carnivores** (animals that eat meat, e.g. tigers) are specially **adapted** to their diets:

RUMINANTS	CARNIVORES
Long digestive tract with a large, modified stomach. (Digesting cellulose is a complicated process that involves different stages.)	**Short** digestive tract with a simple stomach.
Continually **growing** teeth (to stop them wearing down), including **large, ridged molars** for grinding. The canines are only small.	Teeth that **stop growing**, including **pointed molars** for cutting, large, **sharp canines** for gripping live prey, and **sharp incisors** in pairs for biting.
Jaw which moves from **side to side** for grinding.	Jaw which moves **up and down** for cutting and chewing flesh.

Saprophytes Digest Food Outside Their Bodies

Saprophytes feed on **dead organic matter** and **wastes** like faeces.

1) **Rhizopus** is a saprophytic **fungus**, which feeds on starch and other organic materials in plants.
2) Rhizopus fungi have thread-like cells called **hyphae** to penetrate the food.
3) **Enzymes** are released from the hyphae to digest the food surrounding them. Large insoluble molecules are broken down by the enzymes, giving smaller soluble products that are then absorbed into the hyphae cells.

Saprophytic Digestion

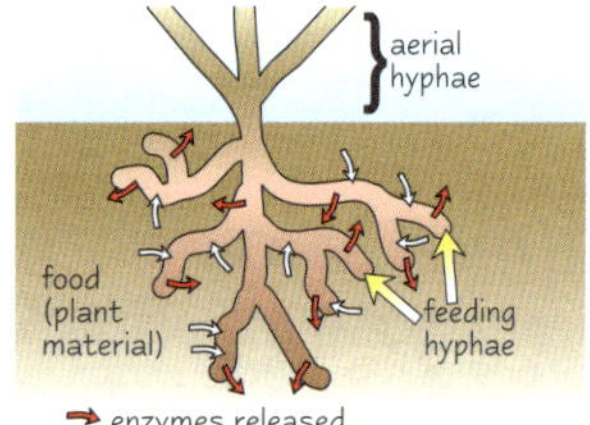

You can see extracellular saprophytic digestion by using the **starch agar assay**. The saprophytic fungi are grown on agar that contains starch. Then the starch agar is flooded with **iodine**, and turns blue-black. But around areas of fungal growth there are **clear zones**. This is because the starch has been digested by the fungi.

Modes of Nutrition

The Tapeworm, Taenia solium, is a Parasite

Parasites live **on** or **in** their **larger host**. The parasite gets its **nutrition** from the host, and this **harms the host**. *Taenia solium* is an example — it's a type of tapeworm that can grow up to 3 metres long. It has a complex life cycle — it starts off its life living in a pig and ends it living in the human gut. Nice.

The *Taenia* tapeworm's body is made up of lots of small sections. One of these sections is called a proglottis, several of them are called proglottides — another one of those funny plurals. The proglottides move further from the head as they mature, and eggs develop inside them.

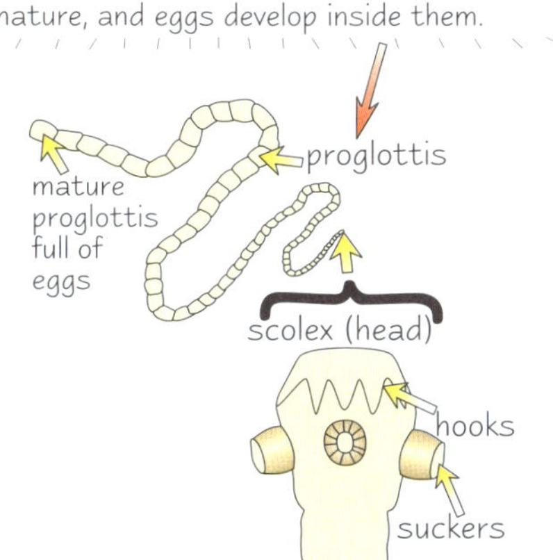

Taenia solium has lots of adaptations to make it an effective parasite:

1) It's **long and thin**, giving it a **large surface area** to absorb more nutrients from the human gut.
2) The long thin shape also prevents the gut getting blocked, because parasites obviously don't want to kill their hosts.
3) *Taenia* has **hooks** and **suckers** on its **scolex** (head part) so it can **grip** onto the intestine wall.
4) There's no digestive system in the tapeworm — the host does the digestion for it, and *Taenia* just absorbs the products of digestion from the gut. Cheeky.
5) The tapeworm's body is covered in a **cuticle**, which prevents its living cells being digested by the host.
6) *Taenia* produces **loads of eggs**, to increase the chances of infecting a new host. The eggs leave in the host's **faeces**.

Mutualism is where Both Partners Benefit

Mutualism is an association between organisms of **different species** where both partners benefit.

An example is between the nitrogen-fixing bacterium, *Rhizobium* (don't confuse with *Rhizopus* at the bottom of p.64) and a leguminous, flowering plant of the family *Papilionaceae*. *Rhizobium* gets into the roots of the plant through its root hairs and causes **root nodules** to develop. It multiplies inside these and **reduces** nitrogen to make **ammonium**. The plant gets a source of ammonium so it can grow, even in poor soils. In return, the *Rhizobium* gets sugars, vitamins and a sheltered habitat.

Another example is the relationship between **ruminants** like cattle and the **microbes** that live in their **rumen** (part of the stomach). The **ruminant** provides **food** and a **warm moist environment** for the microbes. The **microbes digest cellulose** in the grass that the ruminant eats. The ruminant can't digest grass itself because it hasn't got a cellulase enzyme to break down the cellulose found in grass.

Practice Questions

Q1 What mode of nutrition does *Rhizopus* use?

Q2 What is meant by mutualism?

Q3 Define a parasite.

Q4 Draw a diagram of *Taenia solium* and label it.

Exam Questions

Q1 Parasites often have adaptations to help them survive in or on their hosts.
Give four examples of these adaptations, with reference to a named parasite. [5 marks]

Q2 Explain how the saprophyte, *Rhizopus*, is able to digest food outside of its body. [4 marks]

Fungus, tapeworms and faeces — do not attempt to snack while learning...

I know there's a lot of hard-to-remember names and easily-mixed-up terms on this page. But some people have worms inside them that are 3 m long and used to be inside a pig. So count yourself lucky, stop whining and start learning.

Energy Transfer Through an Ecosystem

Two nice, easy pages — loads of it's common sense and some stuff might already be familiar. Just be careful when you're learning all the definitions, because some of the words mean very similar things.

There are **Nine Main Definitions** You **Need To Know**

TERM	MEANING
Biosphere	The part of the Earth's surface and atmosphere inhabited by living things.
Ecosystem	An ecosystem supports life. Nutrients are recycled though an ecosystem, and energy flows through an ecosystem. E.g. a pond, a lawn, a wood. An ecosystem includes both the living and the non-living things there.
Habitat	A place where an organism lives, e.g. a wood pigeon's habitat is a wood.
Producer	Producers make their own food using an external energy source, e.g. light from the Sun. Plants are producers. Organisms that make their own food are also called **autotrophs** (see p.64).
Consumer	Consumers eat other organisms for food and energy. Cows eat a **producer** (grass) so they're **primary** consumers. Humans then eat cows so they're **secondary** consumers. Consumers are also called **heterotrophs** (see p.64).
Decomposer	Decomposers (e.g. bacteria and fungi) break down all the organic material in dead organisms and waste, returning matter to the soil as inorganic material, which is then re-synthesised by the producers. This cycle makes producers and consumers **interdependent**. Decomposers are also called **saprophytes** (see p.64).
Trophic level	A particular **feeding stage** in the food chain, e.g. producer, primary consumer, secondary consumer, etc.
Food chain	A sequence representing the way energy flows from one organism to another.
Food web	A diagram showing all the feeding relationships between all the organisms living in a particular ecosystem. Food webs are made up of many interconnected food chains.

Energy Flows Through **Food Chains**

1) Each organism in a food chain is on a different **trophic level** — it can be a producer, a primary consumer, a secondary consumer, etc.
2) A food chain starts with a **producer**, which uses **energy from the Sun** to synthesise organic material.
3) The **primary consumer** eats the producer, and so on up the food chain.
4) Only about **10%** of the **energy** stored in the organisms at one trophic level passes to the organisms in the next level. Some energy is lost as **heat** from the organisms, and some is lost in their **waste**.

The arrows always point from the organism being eaten to the organism that eats it.

Food Webs are Made Up of Many **Inter-Linked Food Chains**

Most organisms don't eat just one thing — so food chains don't tell the full story. **Food webs** tell you more — for example, this one shows that algae are eaten by three different primary consumers, not just by the water beetle. Like food chains, food webs always start with producers and end with the top consumers.

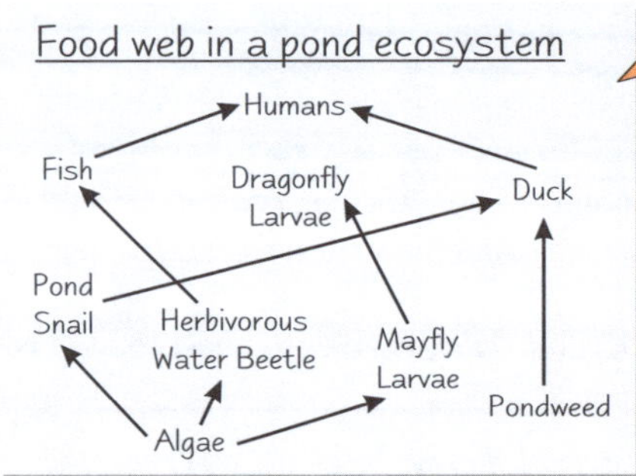

Decomposers are involved at every stage in a food chain or web. They break down dead organisms and return the nutrients in them to the soil so that they can be used by the producers.

Energy Transfer Through an Ecosystem

There are **Three Main Types** of **Ecological Pyramid**

In all these 'pyramids' the **width** of each block is proportional to the figure it's representing.

1) **Pyramids of numbers** show **how many organisms** there are at each trophic level. They're not always pyramid-shaped:

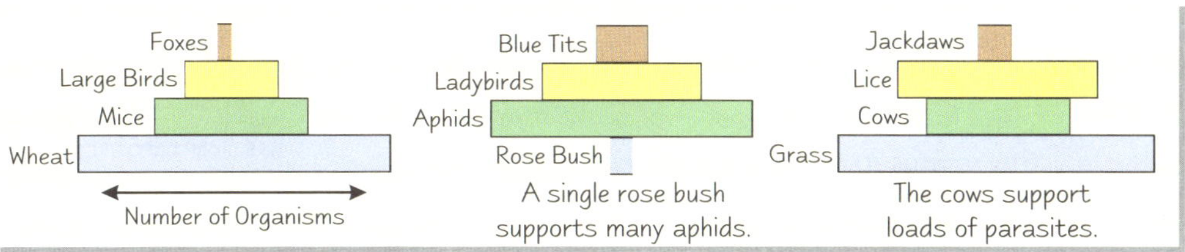

A single rose bush supports many aphids.

The cows support loads of parasites.

2) **Pyramids of biomass** show the **total dry mass** of the organisms (in **kg m^{-2}**) in each trophic level at a particular time. These are **almost always** pyramid-shaped:

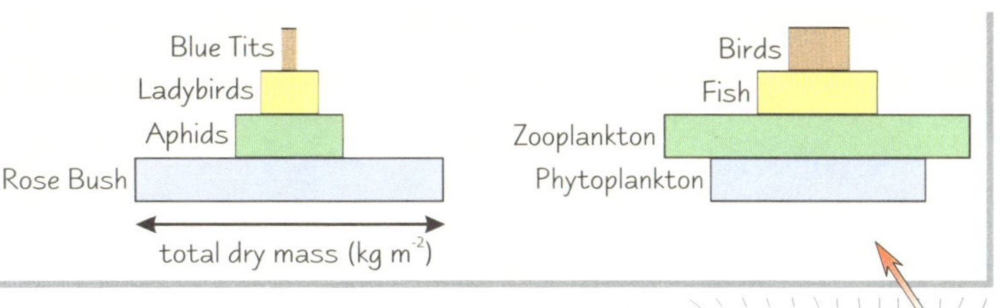

If the producers don't live very long but reproduce loads, their dry mass at a particular time (called the <u>standing crop</u>) won't be very big — even if they support lots of consumers.

3) **Pyramids of energy** show how much energy (in kilojoules per squared metre per year — **kJ m^{-2} y^{-1}**) is passed on from one trophic level to the next. It's never all passed on, so they're **always pyramid-shaped**.

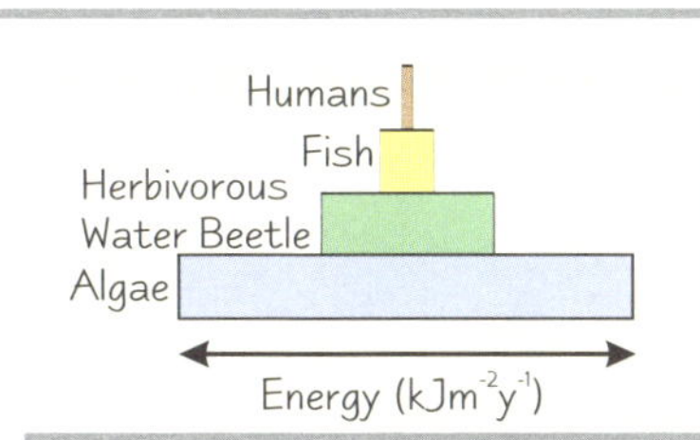

Productivity is About **How Much Food** Organisms **Produce**

1) The **productivity** of a producer is how efficiently it uses energy (e.g. light) to make new tissues — the better it makes tissue, the more food it makes for primary consumers and the more food there is in the ecosystem.
2) **Gross primary productivity** (GPP) is the **total amount** of energy absorbed by the producers **per unit area per year** and used to make new tissues. The units are **kJ m^{-2} y^{-1}**
3) The **net primary productivity** (NPP) is the **energy** that's **available** for the **next trophic level**.
4) So:

NPP = GPP – energy used in respiration or lost as waste by the producers.

Practice Questions

Q1 What is an ecosystem?

Q2 What is meant by the term 'habitat'?

Q3 Which type of ecological pyramid is always pyramid shaped?

Q4 What is the relationship between NPP and GPP?

Exam Questions

Q1 Pyramids of biomass are not always pyramid shaped. Explain why. [2 marks]

Q2 Approximately 1% of the sun's energy is incorporated into the producers. Only 10% of the energy in one trophic level passes to the organisms in the next trophic level. If 10,000 arbitrary units of energy are given out by the sun, how many of these arbitrary units will be incorporated into the secondary consumers? [3 marks]

'Arbitrary' just means that the units could be anything. It doesn't matter what they are, so you don't have to worry about them — which is nice.

Things eat each other — how do they make it so complicated?

A good question, but not one you've got time to ponder now. What you need to do is learn that list of 9 definitions. Some are obvious, like decomposers, but some are more tricky, like the difference between a biosphere and an ecosystem. Don't try to understand the reasons behind the terms, it's a waste of energy. Just get them learnt.

Recycling Nutrients

There are 3 cycles you need to know about — water, carbon and nitrogen. The water cycle's pretty straightforward and the carbon's not too bad, but the nitrogen cycle's evil. But fear not — learn the diagrams and you'll be fine.

Water is Constantly Recycled

1) Water falls as **precipitation** (rain, hail, sleet or snow).
2) Some is used by organisms. This returns to the atmosphere by **evaporation** from respiratory surfaces, or via **transpiration** in plants.
3) Some runs into oceans, rivers and lakes. Water evaporates from these giving **atmospheric water vapour** (clouds).
4) The cycle starts again with precipitation from the clouds.

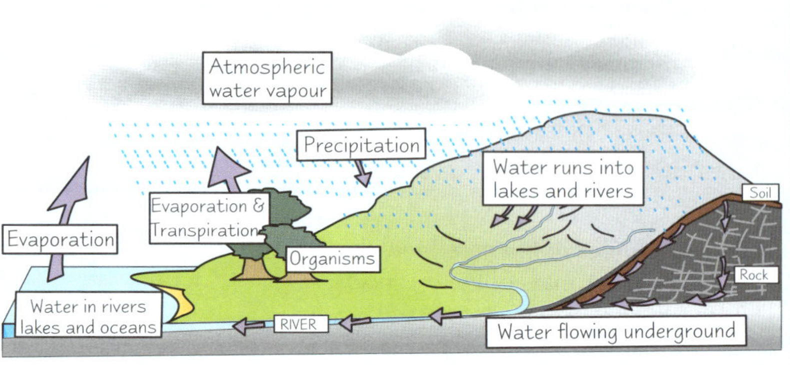

Carbon is Recycled in the Carbon Cycle

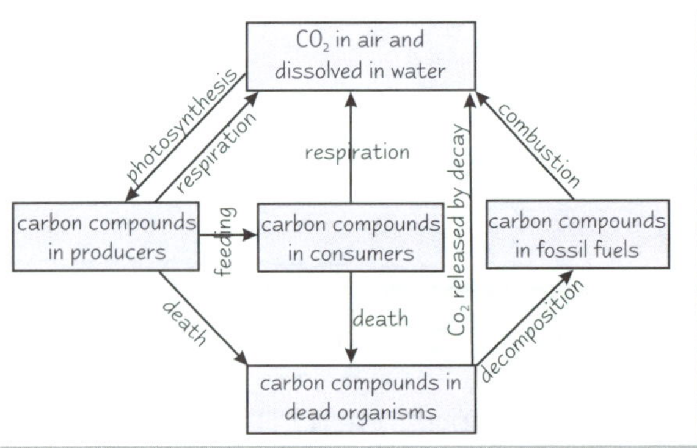

1) Carbon is found as **carbon dioxide gas** (CO_2) in the atmosphere and dissolved in water.
2) CO_2 from the atmosphere is absorbed by **green plants** and used for **photosynthesis** to make carbohydrates. Plants are known as **carbon sinks**, because lots of carbon is locked up in their tissues.
3) These carbohydrates **enter consumers** when they **feed** on the plants and on each other.
4) When organisms **respire**, CO_2 returns to the atmosphere.
5) **Burning fossil fuels** also returns CO_2 to the atmosphere.
6) **Decomposers** break down dead organic matter, releasing CO_2 back into the atmosphere.
7) Carbon follows another major recycling pathway in the **sea**. Marine organisms make **shells** using **carbonates**, which contain carbon. When they die, the shells fall to the ocean floor and eventually form **limestone rocks**. The carbon in these rocks returns to the atmosphere as CO_2 during **volcanic eruptions** or **weathering**.

Microorganisms are Important in the Nitrogen Cycle

Don't let the diagram on the next page scare you — it's not that bad once you start learning it. Honest. Basically, **all living things** need **nitrogen** to make molecules like **proteins**, **ATP** and **DNA**. Plants can only absorb nitrogen in the form of **ammonium** or **nitrate ions**, while **consumers** get their nitrogen by eating other organisms. Here's a list of the facts to help you understand the diagram on page 69:

1) Some of the **ammonium** and **nitrate ions** used by plants are in the soil because of **nitrogen fixation**. This is when **atmospheric nitrogen** is converted to ammonium or nitrate ions by **nitrogen fixing bacteria**. Nitrogen fixing bacteria can live freely in the soil (**e.g. *Azotobacter***) or **symbiotically** in the root nodules of leguminous plants (**e.g. *Rhizobium***).
2) Energy from **lightning** also helps turn atmospheric nitrogen into nitrates.
3) The industrial **Haber process** uses atmospheric nitrogen to make nitrate and ammonia fertilisers. Using these fertilisers increases the concentration of ammonium and nitrate ions in the soil.
4) **Decomposers** turn nitrogen-containing compounds in **dead organisms** into **ammonium ions**. These are then converted to **nitrite** by ***Nitrosomonas*** bacteria, and the nitrite's converted to **nitrate** by ***Nitrobacter*** bacteria. The ammonium to nitrate process is called **nitrification**.
5) **Denitrification** is when nitrates are turned back into atmospheric nitrogen by bacteria like ***Thiobacillus*** and ***Pseudomonas***. This only happens in the absence of oxygen — e.g. in waterlogged soil (because these bacteria need **anaerobic** conditions).

Recycling Nutrients

The Nitrogen Cycle Diagram is Scary but Important

OK, here goes. All those arrows and boxes look alarming, but actually it's pretty straightforward, once you understand the processes that are going on. Link it up to the explanation on p.68 and it'll soon start to make sense.

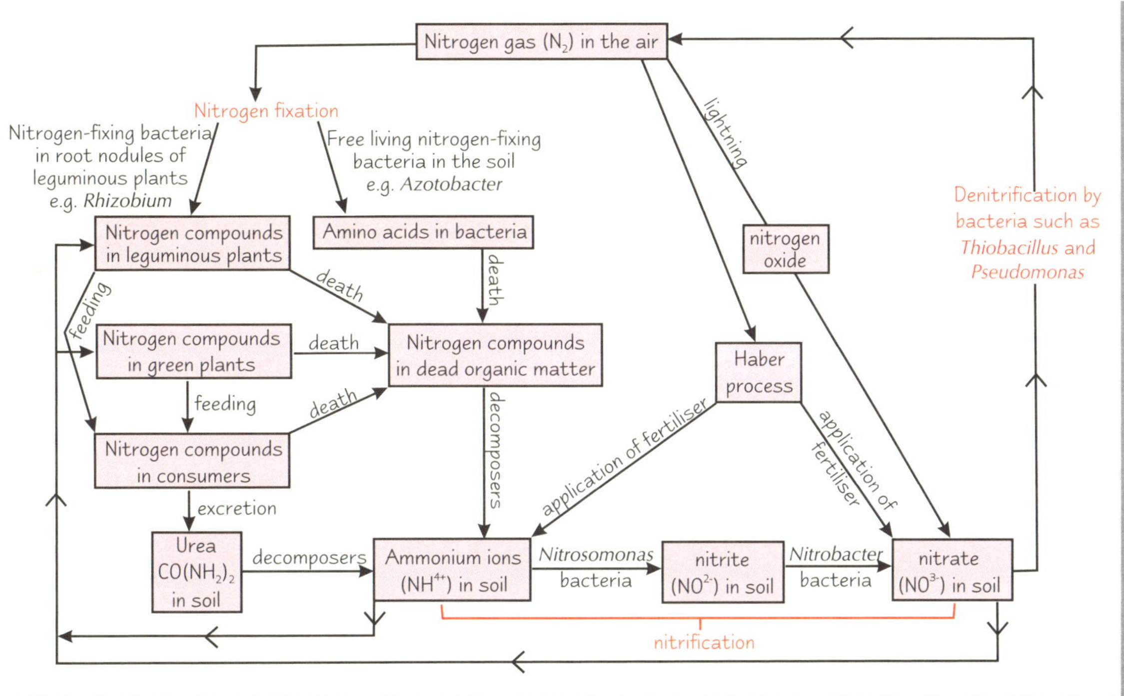

Human Activities Can Disrupt the Carbon and Nitrogen Cycles

1) When **fossil fuels** are burnt they release **CO_2**.
2) Increased levels of CO_2 in the atmosphere are linked to the **Greenhouse Effect**. CO_2 is good at **absorbing heat radiation**, so the more CO_2 in the atmosphere, the hotter it gets. If the Earth's temperature gets too high, it could cause **changes** in **weather patterns** — so there'd be droughts and floods.
3) **Plants** remove CO_2 from the atmosphere. So **deforestation** increases CO_2 levels. When trees are cut down to make room for farmland, they're often **burned**, which releases even more CO_2.
4) **Fertilisers** containing ammonium and nitrates are added to soils. Sometimes, they seep out into lakes and rivers. This causes **eutrophication**, which ends up in the mass death of organisms in the water (see p.71 for more on eutrophication).

Practice Questions

Q1 Describe how water returns to the atmosphere in the water cycle.

Q2 Why are plants known as 'carbon sinks'?.

Q3 What role does lightning play in the nitrogen cycle?

Q4 In what form do plants absorb nitrogen from the soil?

Exam Questions

Q1 Give an account of the nitrogen cycle, highlighting the importance of microbes. [10 marks]

Q2 Name two processes that increase atmospheric carbon dioxide in the normal carbon cycle. [2 marks]

Look at the state of that nitrogen cycle — I want my mum...

It looks scary, I won't deny it. But once you start it's really not hard to learn. The best way to learn the cycles is to try and draw the diagrams (no peeking), see which bits you missed, then try again. When you've managed it perfectly three times in a row it's probably in your head for good, but come back to it now and again just to check.

Human Influences on the Environment

Humans can really hurt the environment. Luckily, stuff can be done to limit our nastiness and — added bonus — if you know about it you'll do better in your exams.

Fossil Fuels are Non-Renewable, but there are Renewable Alternatives

Fossil fuels like **coal** and **oil** are burnt to provide energy (e.g. most power stations are coal-fired). Fossil fuels are **non-renewable** resources because they take **millions of years** to form. Supplies of fossil fuels will **run out** one day — they're being used up faster than they can be replaced. **Alternatives** to fossil fuels include:

1) **Renewable** energy resources like wind, wave and solar energy.
2) Renewable **biomass** energy resources, like wood, gasohol and biogas.

- **Wood** is the main biomass energy resource. Fast-growing trees like **poplar** and **willow** can be **logged sustainably**. The wood can be burnt for heat energy directly, or it can be used to heat water. This creates **steam**, which can turn **turbines** to generate **electricity**.
- **Gasohol** — can be used in **engines** instead of petrol and is a cleaner fuel. It's produced from **ethanol**, when sugar is **fermented** by yeast.
- **Biogas** — is mostly **methane**, and is produced during **anaerobic fermentation** of organic agricultural and domestic waste by bacteria. The methane burns to produce carbon dioxide and water, without any other pollutants.

Deforestation Damages the Environment

Deforestation can lead to changes in the **climate**, **soil erosion**, and **reduction** in the number of **species** in an ecosystem.

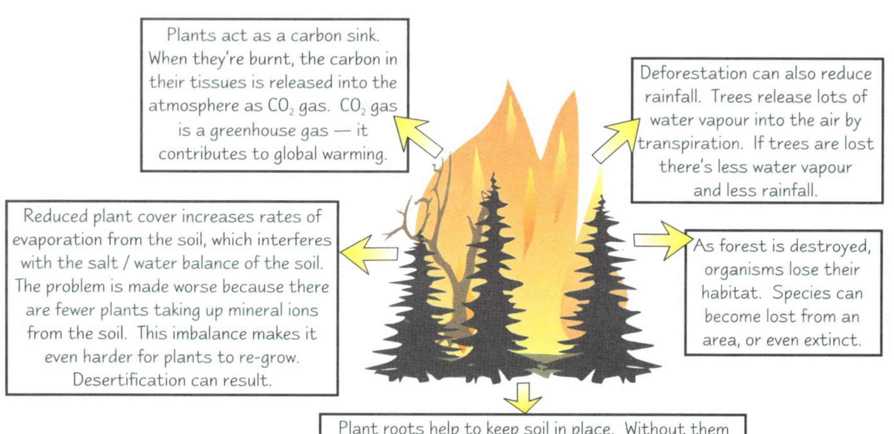

Burning Fossil Fuels Causes Global Warming and Acid Rain

1) CO_2 and other **greenhouse gases** form a **layer** around the Earth. The **Sun's radiation** has a **short wavelength**, and it passes easily through the layer. The radiation from the Sun warms the Earth. As the **Earth** warms up, it **radiates heat** (infrared radiation) from its surface. This radiation has a **longer wavelength**, and gets **trapped** by the greenhouse gases. Because the heat doesn't all escape into space, the Earth gets very slowly warmer and warmer.
2) **Sulphur dioxide** and **oxides of nitrogen** are also produced when fossil fuels are burnt. These gases dissolve in water vapour to make **acids**, which fall to Earth in acid rain or snow. Acid rain can harm plants and animals, mostly by interfering with their **metabolic reactions**.

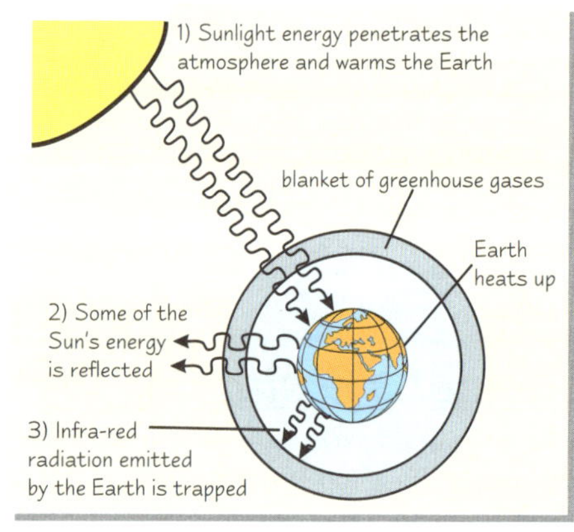

Human Influences on the Environment

Raw Sewage and *Fertilisers* Cause *Eutrophication*

If **raw sewage** or **fertilisers** get into rivers and lakes, they increase the levels of **nitrate** and **phosphate** in them. This leads to **eutrophication**.

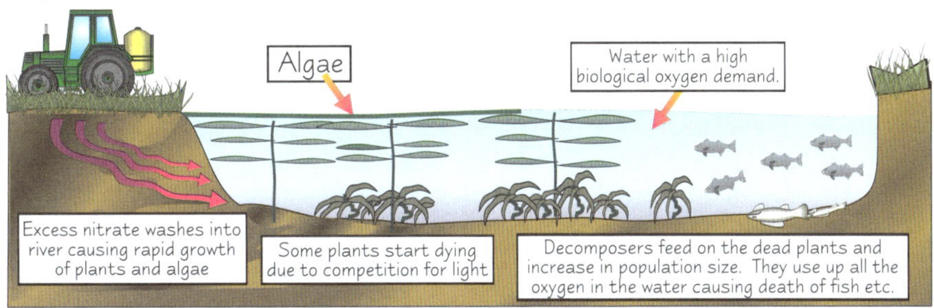

1) Growth of **algae** and other plants increases due to the higher levels of nitrates and phosphates. These **algal blooms** mean that there are so many producers that they become **overcrowded**, and start to die faster than they're eaten.
2) The dead producers provide lots of food for **decomposers**, which flourish. They use up a lot of the oxygen in the water, increasing its **biological oxygen demand** (BOD). With raw sewage, which is full of bacteria anyway, the BOD of the water is even greater.
3) The lack of oxygen leads to the **death** of fish, plants and other aerobic organisms.
4) This whole process is called **eutrophication**.

Conservation Aims to *Protect Natural Resources*

1) An important aim of conservation is to maintain **species diversity**. If one organism becomes extinct it affects other organisms in the ecosystem — because a food supply or predator is removed.
2) **Sustainable agriculture** can help maintain species diversity by allowing the environment time to recover between crops. **Sustainable forestry** involves planting new trees to replace ones that are cut down.
3) Having a **variety** of **habitats** in an area encourages species diversity. This is hard if the area is being used to grow one particular crop or tree, but things like allowing old hedgerows to remain can really improve species diversity.
4) **EU legislation** has been drawn up to control air and water quality, protect the environment and avoid over-exploitation of natural resources.

Practice Questions

Q1 Name three types of renewable energy resources.

Q2 How is gasohol produced?

Q3 Name two pollution problems that can be caused by burning fossil fuels.

Q4 Name three problems caused by deforestation.

Ha, ha, ha, ha
Stayin' alive, stayin' alive

Exam Question

Q1 Fertilisers can be damaging to lakes and rivers. Explain why. [8 marks]

I was told gases are better out than in — obviously not for fossil fuels...

You know the phrase 'It ain't over 'til the fat pig sings'? Well, there he goes, look, he's singing away ever so happily. That means the book is officially over. Which is something that I'm rather excited about. I'm now off to play on my dance mat, and eat chocolate brownies. Mmmm. Wait, where are you going? Your revision isn't over yet! Come back! Revise!

Answers

Section 1 — Biological Molecules

Page 3 — Water

1 *Maximum of 12 marks available, from any of the 14 mark points listed.*
Water molecules have two hydrogen atoms and one oxygen atom ***[1 mark]****.*
The hydrogen and oxygen are joined by covalent bonds / sharing electrons ***[1 mark]****.*
Water molecules are polar ***[1 mark]****.*
Polarity leads to the formation of hydrogen bonds between water molecules ***[1 mark]****.*
Water is a solvent ***[1 mark]****.*
Water's polar nature allows water to dissolve polar solutes ***[1 mark]****.*
Water transports substances ***[1 mark]****.*
Substances are transported more easily when dissolved ***[1 mark]****.*
Water has a high specific heat capacity ***[1 mark]****.*
This means it's difficult to change the temperature of water ***[1 mark]****.*
This allows cells to avoid sudden changes in temperature ***[1 mark]****.*
Water has a high latent heat of evaporation ***[1 mark]****.*
This is due to hydrogen bonding ***[1 mark — but only awarded if hydrogen bonding hasn't already been mentioned]****.*
It means the evaporation of water has a considerable cooling effect ***[1 mark]****.*
Be careful to stick to what the question asks for. The fact that ice floats on water and aquatic habitats have stable temperatures aren't things that happen "in living organisms".

Page 5 — Carbohydrates

1 *Maximum of 7 marks available.*
Glycosidic bonds are formed by condensation reactions ***[1 mark]*** *and broken by hydrolysis reactions* ***[1 mark]****.*
When a glycosidic bond is formed in a condensation reaction, a hydrogen ***[1 mark]*** *from one monosaccharide combines with a hydroxyl / OH group* ***[1 mark]*** *from the other to form a molecule of water* ***[1 mark]****.*
A hydrolysis reaction is the reverse of this ***[1 mark]****, with a molecule of water being used up to split the monosaccharide molecules apart* ***[1 mark]****.*
The last 5 marks for this question could be obtained by a diagram showing the reaction, using structural formulae.

2 *Maximum of 10 marks available.*
Glycogen is a chain of alpha glucose molecules ***[1 mark]*** *whereas cellulose is a chain of beta glucose molecules* ***[1 mark]****.*
Glycogen's chain is compact and very branched ***[1 mark]*** *whereas cellulose's chain is long, straight and unbranched* ***[1 mark]*** *and these chains are bonded together to form strong fibres* ***[1 mark]****.*
Glycogen's structure makes it a good food store in animals ***[1 mark]****. The branches allow enzymes to access the glycosidic bonds to break the food store down quickly* ***[1 mark]****.*
Cellulose's structure makes it a good supporting structure in cell walls ***[1 mark]****. The fibres provide strength* ***[1 mark]****.*
The function is helped by the fact that the cell doesn't have any enzymes that can break down beta bonds ***[1 mark]****.*
This question is worth 10 marks so you need to include at least 10 relevant points to score full marks. Also, the question asks you to compare and contrast, so make sure you don't just describe glycogen and cellulose totally separately from each other.

Page 7 — Lipids

1 *Maximum of 10 marks available.*
Up to 5 marks for correctly naming the 5 functions.
Up to 5 marks for explaining the relevant features.
Energy store ***[1 mark]*** *— lipids contain a lot of energy per gram* ***[1 mark]****.*
Insulation ***[1 mark]*** *— the lipid layer under the skin doesn't have an extensive blood supply so heat isn't lost from it* ***[1 mark]****.*
Buoyancy ***[1 mark]*** *— the lipid layer is less dense than muscle and bone* ***[1 mark]****.*
Protection ***[1 mark]*** *— the fat layer under the skin and around internal organs acts like a cushion to prevent damage from any blows* ***[1 mark]****.*
Waterproofing ***[1 mark]*** *— lipids don't mix / dissolve in water, so water can't penetrate a lipid layer* ***[1 mark]****.*

2 *Maximum of 8 marks available.*
Triglycerides are made from a glycerol molecule ***[1 mark]*** *and three molecules of fatty acids* ***[1 mark]****.*
They are formed by condensation reactions ***[1 mark]****.*
These reactions result in the formation of ester bonds ***[1 mark]*** *between the fatty acid and glycerol molecules, with the production of a molecule of water for each fatty acid added* ***[1 mark]****.*
Triglycerides are broken up by hydrolysis reactions ***[1 mark]****, which are the reverse of condensation reactions* ***[1 mark]****, with one molecule of water being added for each fatty acid that's released* ***[1 mark]****.*
It would be possible to get all the marks in this question by using labelled diagrams, as long as all the points listed have been illustrated.

3 *Maximum of 8 marks available.*
A triglyceride consists of glycerol ***[1 mark]*** *and three fatty acid molecules* ***[1 mark]****.*
A phospholipid has the same basic structure, but one of the fatty acids is replaced by a phosphate group ***[1 mark]****.*
Triglycerides are hydrophobic / repel water ***[1 mark]****.*
This is a property of the hydrocarbon chains that are part of the fatty acid molecules ***[1 mark]****.*
The phosphate group in a phospholipid is hydrophilic / attracts water ***[1 mark]****, because it's ionised* ***[1 mark]****.*
This means that the phospholipid has a hydrophilic 'head' and a hydrophobic 'tail' ***[1 mark]****.*
The words 'head' and 'tail' are not essential, as long as you've got across the idea that the molecule is partly hydrophobic and partly hydrophilic.

Page 9 — Proteins

1 *Maximum of 9 marks available.*
Proteins are made from amino acids ***[1 mark]****.*
The amino acids are joined together in a long (polypeptide) chain ***[1 mark]****.*
The sequence of amino acids is the protein's primary structure ***[1 mark]****.*
The amino acid chain / polypeptide coils in a certain way ***[1 mark]****.*
The way it's coiled is the protein's secondary structure ***[1 mark]****.*
The coiled chain is itself folded into a specific shape ***[1 mark]****.*
This is the protein's tertiary structure ***[1 mark]****.*
Different polypeptide chains can be joined together in the protein molecule ***[1 mark]****.*
This is the quaternary structure of the protein ***[1 mark]****.*
The question specifically states that you don't need to describe the chemical nature of the bonds in a protein. So, even if you name them, don't go into chemical details — you won't get any credit for it.

2 *Maximum of 6 marks available.*
Collagen is a fibrous protein ***[1 mark]****.*
For this mark, mentioning that the molecule is fibrous is essential.
It forms supportive tissues in the body ***[1 mark]****.*
Collagen consists of three polypeptide chains ***[1 mark]****.*
These chains form a triple helix ***[1 mark]****.*
The chains are tightly coiled together ***[1 mark]****.*
The tightly coiled chains provide strength to the structure ***[1 mark]****.*
Minerals can bind to the collagen chain ***[1 mark]****.*
This makes it more rigid ***[1 mark]****.*
8 marks are listed, but the mark is given out of 6. This is common in longer exam questions. You would have to be a bit of a mind-reader to hit every mark the examiner thinks of, so to make it fair, there are more mark points than marks. You can only count a maximum of 6, though.

Page 11— Structure of DNA and The Genetic Code

1 *Maximum of 2 marks available.*
The long length and coiled nature of DNA molecules allows the storage of vast quantities of information ***[1 mark]****.*
You only get the mark here if you've mentioned the length and the coiled nature.
Good at replicating itself because of the two strands being paired ***[1 mark]****.*
The question is worth two marks so you need to mention at least two things.

2 *Maximum of 5 marks available.*

Answers

*Nucleotides are joined by condensation reactions **[1 mark]**.*
*This happens between the phosphate group and the sugar of the next nucleotide **[1 mark]**.*
*The DNA strands join through hydrogen bonds **[1 mark]** between the base pairs **[1 mark]**.*
*The final mark is given for at least one accurate diagram showing at least one of the above processes **[1 mark]**.*
As the question asks for a diagram make sure you do at least one.

Page 13 — DNA Self-Replication and RNA

1 *Maximum of 6 marks available.*
*DNA strands uncoil and separate **[1 mark]**.*
*Individual free DNA nucleotides pair up with their complementary bases on the template strand **[1 mark]**.*
*DNA polymerase joins the individual nucleotides together **[1 mark]**.*
Students often forget to mention this enzyme in their answers.
*Hydrogen bonds then form between the bases on each strand **[1 mark].***
*Two identical DNA molecules are produced **[1 mark]**.*
*Each of the new molecules contains a single strand from the original DNA molecule and a single new strand **[1 mark]**.*

Page 15 — Protein Synthesis

1 *Maximum of 2 marks available.*
*A codon is a triplet of bases found on a mRNA molecule **[1 mark]**.*
*An anticodon is a triplet of bases found on a tRNA molecule **[1 mark]**.*
The examiner only expects a brief answer to a two-mark question.

2 *Maximum of 10 marks available.*
*Transcription happens inside the nucleus and translation outside in the cytoplasm **[1 mark]**.*
*A section of DNA is uncoiled by breaking hydrogen bonds **[1 mark]**.*
*mRNA makes a copy of an uncoiled section of DNA **[1 mark]**.*
*The mRNA travels outside the nucleus to a ribosome **[1 mark]**.*
*The codons on the mRNA are paired with anticodons on a molecule of tRNA **[1 mark]**.*
*The tRNA molecules are carrying amino acids **[1 mark]** which line up and are joined by peptide bonds with an enzyme **[1 mark]**.*
You must mention that an enzyme is involved to get this mark.
*Specific codons / base triplets code for specific amino acids **[1 mark]**.*
*A polypeptide chain is formed **[1 mark]** — the primary structure of a protein **[1 mark]**.*

Section 2 — Enzymes

Page 17 — Action of Enzymes

1 *Maximum of 2 marks available.*
*Chemical X is an enzyme inhibitor **[1 mark]***
*Reason — it reduces an enzyme controlled reaction **[1 mark]***

Page 19 — Factors that Affect Enzyme Activity / Industrial Production of Enzymes

1 *Maximum of 8 marks available, from any of the 9 points below.*
*If the solution is too cold, the enzyme will work very slowly **[1 mark]**.*
*This is because, at low temperatures, the molecules move slowly and collisions are less likely between enzyme and substrate molecules **[1 mark]**.*
The marks above could also be obtained by giving the reverse argument — a higher temperature is best to use because the molecules will move fast enough to give a reasonable chance of collisions.
*If the temperature gets too high, the reaction will stop **[1 mark]**.*
*This is because the enzyme is denatured **[1 mark]** — the active site changes and will no longer fit the substrate **[1 mark]**.*
*Denaturation is caused by increased vibration breaking bonds in the enzyme **[1 mark]**.*
*Enzymes have an optimum pH **[1 mark]**.*
*pH values too far from the optimum cause denaturation **[1 mark]**.*
Explanation of denaturation here will get a mark only if it hasn't been explained earlier.
*Denaturation by pH is caused by disruption of ionic bonds, which destabilises the enzyme's tertiary structure **[1 mark]**.*

2 a) *Maximum of 2 marks available.*
*Immobilised enzymes aren't free in solution **[1 mark]**.*
*Instead they're absorbed on / trapped in / encapsulated in an inert (non-reactive) material **[1 mark]**.*
It might seem like these two points are saying the same thing but they are actually different, so you have to make sure you say both of them.

b) *Maximum of 2 marks available. Any 2 of these:*
*Does not contaminate the product **[1 mark]**.*
*Can be easily recovered and used again **[1 mark]**.*
*More stable **[1 mark]**.*

Section 3 — Cellular Organisation

Page 21 — Cell Organisation

1 *Maximum of 12 marks available.*
Up to 6 marks for correctly naming each leaf tissue.
Up to 6 marks for explaining each tissue's adaptation:
*Lower epidermis **[1 mark]** — has stomata which allow carbon dioxide and oxygen in and out **[1 mark]**.*
*Spongy mesophyll **[1 mark]** — has air spaces which allow gases to circulate **[1 mark]**.*
*Palisade mesophyll **[1 mark]** — has many chloroplasts to absorb sunlight **[1 mark]**.*
*Upper epidermis **[1 mark]** — waterproof to keep water in **[1 mark]**.*
*Xylem **[1 mark]** — delivers water to the leaf **[1 mark]**.*
*Phloem **[1 mark]** — transports sugars away from the leaf **[1 mark]**.*

Page 23 — Electron and Light Microscopy

1 *Maximum of 6 marks available.*
***Advantages**: Greater resolution **[1 mark]**.*
*More detail / internal structure of organelles can be seen **[1 mark]**.*
***Disadvantages**: Electron microscopes can't be used to study living tissues **[1 mark]**.*
*Natural colours can't be seen **[1 mark]**. They aren't portable **[1 mark]**. They are expensive **[1 mark]**.*

Page 25 — Functions of Eukaryotic Organelles

1 *Maximum of 9 marks available.*
You could have written about any three of the four organelles given below.
***Mitochondria** **[1 mark]** — Large numbers of mitochondria would indicate that the cell used a lot of energy **[1 mark]**, because mitochondria are the site of (aerobic) respiration, which releases energy **[1 mark]**.*
***Chloroplasts** **[1 mark]** — Large numbers of chloroplasts would be seen in cells that are involved in photosynthesis **[1 mark]** because the chloroplasts contain chlorophyll, which absorbs light for photosynthesis **[1 mark]**.*
Rough endoplasmic reticulum** **[1 mark]** — You find a lot of RER in cells that produce a lot of protein **[1 mark]** because the RER transports protein made in the attached ribosomes **[1 mark]
***Lysosomes** **[1 mark]** — Found in cells that are old / destroy other cells **[1 mark]** because lysosomes contain digestive enzymes that can break down cells **[1 mark]**.*
There are 9 marks for this question and 3 organelles have to be mentioned. You get a mark for mentioning each of the correct organelles, then you need to give 2 pieces of relevant information for each one.

2 a) *Maximum of 2 marks available.*
*i) mitochondrion **[1 mark]***
*ii) Golgi apparatus **[1 mark]***

b) *Maximum of 2 marks available:*
*The function of the mitochondrion is to be the site of (aerobic) respiration / provide energy **[1 mark]**.*
*The function of the Golgi apparatus is to package materials made in the cell / to make lysosomes **[1 mark]**.*
The question doesn't ask you to give the reasons why you identified the organelles as you did, so don't waste time writing your reasons down.

Page 27 — The Cell Membrane Structure

1 a) *Maximum of 1 mark available.*
In a triglyceride there are three fatty acids attached to a molecule of

Answers

*glycerol but in a phospholipid one of the fatty acids is replaced by a phosphate group **[1 mark]**.*

b) *Maximum of 2 marks available.*
*Phospholipids are arranged in a double layer / bilayer with fatty acid tails on the inside **[1 mark]**.*
*Fatty acid tails are hydrophobic / non-polar so they prevent the passage of water soluble molecules through the cell membrane **[1 mark]**.*
You might be asked to draw a layer of phospholipid molecules on the surface of a container of water. You should draw the molecules with their hydrophilic phosphate heads in the water and their hydrophobic fatty acid tails sticking up into the air.

2 a) *Maximum of 2 marks available.*
*A is an extrinsic protein / receptor protein **[1 mark]**.*
*B is an intrinsic protein / carrier protein **[1 mark]**.*

b) *Maximum of 2 marks available.*
*Tertiary structure gives the receptor protein a particular shape **[1 mark]***
*Only molecules with a specific shape will fit into (or bind with) the protein **[1 mark]***

Page 31 — Transport Across the Cell Membrane

1 *water potential = solute potential + pressure potential **[2 marks]**.*

2 *Maximum of 3 marks available.*
*The water potential of the hypotonic solution would be higher than the water potential of the cell **[1 mark]**. Therefore, there would be a net movement of water molecules, by osmosis, from the solution into the cell **[1 mark]**. The cell would eventually burst **[1 mark]**.*

3 *Maximum of 5 marks available.*
*The cell surrounds a solid particle with a section of its cell membrane **[1 mark]**. This section of the membrane forms a small vacuole around the particle called a vesicle **[1 mark]**. The vesicle separates from the main membrane and moves into the cell's cytoplasm **[1 mark]**. The contents of the vesicle are digested by enzymes secreted by lysosomes **[1 mark]**. Molecules and ions then diffuse out of the vesicle into the cell's cytoplasm **[1 mark]**.*

Section 4 — The Cell Cycle
Page 33 — The Cell Cycle and Mitosis

1 a) *Maximum of 2 marks available.*
*Interphase **[1 mark]**, during the S or Synthesis stage **[1 mark]**.*

b) *Maximum of 2 marks available:*
*Mitosis **[1 mark]**, during the prophase stage **[1 mark]**.*
Sometimes you can pick up marks even if you're only half right. You might know spindle fibres are formed during mitosis, but you've forgotten which stage. It's worth mentioning mitosis anyway, and then having a guess.

Section 5 — Exchanges With the Environment
Page 34-35 — Nutrient and Gaseous Exchange / Gaseous Exchange in Flowering Plants

1 *Maximum of 4 marks available.*
*Humans are large multicellular organisms **[1 mark]**.*
*The surface area : volume ratio is small in large organisms **[1 mark]**, which makes diffusion too slow **[1 mark]**.*
*Humans need specialised organs with a large enough surface area to keep all their cells supplied with enough oxygen and to remove CO_2 **[1 mark]**.*

2 *Maximum of 6 marks available:*
*Potassium ions are pumped into the guard cells **[1 mark]** by active transport **[1 mark]**.*
*Starch in the chloroplasts is converted into malate **[1 mark]**.*
*This increases the solute concentration / decreases the water potential of the cell sap **[1 mark]**.*
*Water flows into the cell by osmosis **[1 mark]**.*
*Due to the different thicknesses of the guard cell's cell walls, the guard cell bulges outwards, opening the stoma **[1 mark]**.*
Any of the marks above could be obtained by drawing an annotated diagram. Using a diagram could save time in the exam, and make it easier to remember all six points.

Page 37 — Gaseous Exchange in Humans

1 *Maximum of 3 marks available.*
*Tidal volume would go up **[1 mark]** as would the breathing rate. This is because of the increased demand for oxygen for aerobic respiration **[1 mark]** and decrease in pH due to extra CO_2 in the blood **[1 mark]**.*

Page 39 — Digestion and Absorption

1 *Maximum of 4 marks available, from any of the 8 points below.*
*Polypeptides are broken down by peptidases **[1 mark]** to form amino acids **[1 mark]** when peptide bonds are hydrolysed **[1 mark]**. Exopeptidases hydrolyse peptide bonds between amino acids on the outside **[1 mark]** of the polypeptide chain. Endopeptidases hydrolyse peptide bonds between amino acids on the inside **[1 mark]** of the polypeptide chain. Peptidases are released into the acidic conditions of the stomach in gastric juice **[1 mark]** and in the duodenum from pancreatic juice **[1 mark]** and from the epithelial cells lining the small intestine **[1 mark]**.*

2 *Maximum of 3 marks available.*
*Small, soluble products of digestion (e.g. glucose, amino acids, fatty acids) **[1 mark]** are absorbed into the body through microvilli lining the gut wall **[1 mark]**.*
*Absorption is through diffusion, facilitated diffusion and active transport **[1 mark if all 3 methods are mentioned]**.*

Section 6 — Transport Systems
Page 40-41 — Transport in Animals and Plants / Transport in Flowering Plants

1 *Maximum of 13 marks available:*
*Only 1 mark for correctly naming all 4 cell types — vessel elements, tracheids, parenchyma cells, fibre cells **[1 mark]**.*
***Vessels** — Up to 2 marks for correctly naming features, from any of the following: Thickened and lignified walls **[1 mark]**. No cytoplasm / dead cells **[1 mark]**. No end walls **[1 mark]**.*
*Functions — transport of water / mineral ions and support **[1 mark]**. Both functions needed for mark.*
***Tracheids** — Thickened and lignified walls **[1 mark]**.*
*No cytoplasm/dead cells **[1 mark]***
*Functions — transport of water and support **[1 mark]**.*
Both functions needed for mark.
***Parenchyma** — Living cells **[1 mark]**. Unthickened walls **[1 mark]**.*
*Function — food storage/radial transport **[1 mark]**.*
***Fibres** — Very thick walls **[1 mark]**. Narrow lumen **[1 mark]**.*
*Function — support **[1 mark]**.*
Pay attention to how much you need to know about each type of cell. There's more to remember about the structure of vessels than about the structure of fibres, so make sure you spend more time learning it.

2 *Maximum of 5 marks available:*
*Sieve tube elements are joined end-to-end to form sieve tubes **[1 mark]**.*
*The 'sieve' parts are the end walls, which have lots of holes in them **[1 mark]**, connecting the cytoplasm of adjacent cells **[1 mark]**.*
*They have no nucleus **[1 mark]**.*
*They have only a very thin layer of cytoplasm without many organelles **[1 mark]**.*

Page 43 — Movement of Water

1 *Maximum of 6 marks available:*
*Water passes through the cell walls en route to the xylem **[1 mark]**.*
*This is the apoplast pathway **[1 mark]**.*
*Plasmodesmata connect the cytoplasm of adjacent cells so water can travel from cell to cell through them **[1 mark]**.*
*This is the symplast pathway **[1 mark]**.*
*Water travels from the vacuole of one cell into the vacuole of the next by osmosis **[1 mark]**.*
*This is the vacuolar pathway **[1 mark]**.*

2 *Maximum of 4 marks available:*
*Loss of water from the leaves due to transpiration pulls more water in from xylem **[1 mark]**.*

Answers

*There are cohesive forces between water molecules **[1 mark]**.*
*These cause water to be pulled up the xylem **[1 mark]**.*
*Removing leaves means no transpiration occurs, so no water is pulled up the xylem **[1 mark]**.*
It's pretty obvious (because there are 4 marks to get) that it's not enough just to say removing the leaves stops transpiration. You also need to explain why transpiration is so important in moving water through the xylem. It's always worth checking how many marks a question is worth — this gives you a clue about how many details you need to include.

Page 45 — Circulation, the Heart and the Cardiac Cycle

1 *Maximum of 3 marks available:*
*Pressure increases in atria during atrial systole and in ventricles during ventricular systole **[1 mark]**.*
*Pressure decreases in atria during atrial diastole and in the ventricles during ventricular diastole **[1 mark]**.*
*There is always more pressure on the left side of the heart due to extra muscle tissue producing more force **[1 mark]**.*
This question doesn't ask you to describe the cardiac cycle — it specifically asks you to describe the pressure changes in diastole and systole. Make sure you mention both atria and ventricles in your answer.

2 *Maximum of 6 marks available:*
*The valves only open one way **[1 mark]**.*
*Whether they open or close depends on the relative pressure of the heart chambers **[1 mark]**.*
*If the pressure is greater behind a valve (i.e. there's blood in the chamber behind it) **[1 mark]**, it's forced open, to let the blood travel in the right direction **[1 mark]**.*
*When the blood goes through the valve, the pressure is greater above the valve **[1 mark]**, which forces it shut, preventing blood from flowing back into the chamber **[1 mark]**.*
Here you need to explain how valves function in relation to blood flow, rather than just in relation to relative pressures.

Page 46-47 — Blood Vessels / Control of Heartbeat

1 *Maximum of 6 marks available:*
*Arteries are thick-walled **[1 mark]**, muscular **[1 mark]** and have elastic tissue in the walls **[1 mark]**.*
*Veins are wider than equivalent arteries **[1 mark]**, with very little elastic or muscle tissue **[1 mark]**. Veins contain valves **[1 mark]**.*

2 *Maximum of 6 marks available:*
*The SAN produces a stimulus **[1 mark]** which causes the left and right atrial muscle to contract **[1 mark]**.*
*The AVN picks up the stimulus **[1 mark]** and sends impulses through the Bundle of His **[1 mark]** and on to the Purkyne tissue in the ventricle walls **[1 mark]**, which enable the left and right ventricles to contract **[1 mark]**.*
Examiners hate the use of 'messages' or 'signals' for nerve impulses — use the correct terms so you don't throw away easy marks.

Page 49 — Blood and Tissue Fluid

1 *Maximum of 5 marks available:*
*Tissue fluid moves out of capillaries by pressure filtration / hydrostatic pressure **[1 mark]**.*
*At the arteriole end, pressure in capillary beds is greater than pressure in tissue fluid outside capillaries **[1 mark]**.*
*This means fluid from blood is forced out of the capillaries **[1 mark]**.*
*Fluid loss means the water potential of blood capillaries is lower than that of tissue fluid **[1 mark]**.*
*So fluid moves into the capillaries at the venule end by osmosis **[1 mark]**.*

2 *Maximum of 3 marks available:*
*Phagocytes engulf pathogens / microorganisms and digest them **[1 mark]**.*
*They contain many lysosomes to aid digestion **[1 mark]**.*
*They have elongated nuclei, and fluid cytoplasm so they can squeeze through gaps to reach the site of infection **[1 mark]**.*
Exam questions relating to structure and function are a favourite. Find a way (cartoons, poems, tables) to remember the name, structure and function all together to avoid confusion.

Page 51 — Haemoglobin and Oxygen Transport

1 *Maximum of 3 marks available:*
*The foetus relies on oxygen diffused from the mother's blood **[1 mark]**.*
*If the haemoglobin of the foetal and mother's blood had the same affinity for oxygen they'd be competing for oxygen and diffusion wouldn't happen **[1 mark]**.*
*So foetal haemoglobin has a higher affinity for oxygen than its mother's blood **[1 mark]**.*

2 *Maximum of 4 marks available:*
*When tissues are active, cells respire more quickly **[1 mark]**.*
*Oxygen is used up faster, creating a low pO_2 **[1 mark]**.*
*Also, more CO_2 is released by cells, giving a higher pCO_2 **[1 mark]**.*
*Both factors mean oxygen unloading from oxyhaemoglobin is increased **[1 mark]**.*
Make sure you mention both pO_2 and pCO_2 in questions about the Bohr effect.

Section 7 — Adaptations to the Environment

Page 53 — Environmental Adaptations

1 *Maximum of 3 marks available:*
*Behavioural — buries head in mud with tail pointing upwards. Wiggles tail about to circulate oxygenated water **[1 mark]**.*
*Physiological — blood contains haemoglobin to help it absorb oxygen **[1 mark]**.*
*Structural — capillaries very close to thin surface at the tail end so that oxygen can diffuse into the worm through the surface **[1 mark]**.*

2 *Maximum of 5 marks available:*
*10 cm^3l^{-1} — fast-flowing river has more oxygen than a pond **[1 mark]** because the movement of water incorporates oxygen **[1 mark]**.*
*210 cm^3l^{-1} — terrestrial environment **[1 mark]**. This is 21% oxygen, which is the concentration of oxygen in the atmosphere.*
*More oxygen in terrestrial environments than aquatic. **[1 mark only available, for either or both of these explanations.]**.*
*5 cm^3l^{-1} — pond **[1 mark]**.*

Section 8 — Sexual Reproduction

Page 55 — Meiosis and Sexual Reproduction

1 a) *Maximum of 3 marks available.*
*A = 46 **[1 mark]**.*
*B = 23 **[1 mark]**.*
*C = 23 **[1 mark]**.*

b) *Maximum of 2 marks available, from any of the points below.*
*Normal body cells have two copies of each chromosome, which they inherit from their parents **[1 mark]**.*
*Gametes have to have half the number of chromosomes so that when fertilisation takes place, the resulting embryo will have the correct diploid number **[1 mark]**.*
*If the gametes had a diploid number, the resulting offspring would have twice the number of chromosomes that it should have **[1 mark]**.*

Page 57 — Sexual Reproduction in Plants

1 a) *Maximum of 4 marks available.*
*A = Pollen tube **[1 mark]**.*
*B = Embryo sac **[1 mark]**.*
*C = Tube nucleus **[1 mark]**.*
*D = Micropyle **[1 mark]**.*

b) *Maximum of 2 marks available.*
*The enzymes digest surrounding cells **[1 mark]**.*
*It makes a path through to the ovary **[1 mark]**.*

2 *Maximum of 2 marks available, from any of the points below:*
*It has long filaments and large anthers that stick out **[1 mark]**.*
*It has a long, feathery / exposed stigma **[1 mark]**.*
*It has no petals **[1 mark]**.*
You wouldn't get a mark for saying it's not scented. You'd be right, but you can't tell that from the diagram. Always read the question properly, and if they ask you to use a diagram make sure you do.

Answers

Page 59 — Reproduction in Humans

1 a) *Maximum of 2 marks available.*
At Y — Primary follicle has developed into a mature Graafian follicle ***[1 mark]****.*
At Z — Graafian follicle ruptures to release the egg into the fallopian tube — ovulation ***[1 mark]****.*

b) *Maximum of 3 marks available, from any of the points below.*
Primary follicle stimulated to develop by the sex hormone FSH ***[1 mark]****.*
Follicle matures into a Graafian follicle, which migrates to the surface of the ovary ***[1 mark]****.*
Luteinising hormone is released into the bloodstream, which causes the follicle to burst ***[1 mark]*** *releasing the egg into the fallopian tube* ***[1 mark]****.*
This is an 'explain' question so needs more detail. To get it right you have to remember that the oocyte's fertilised in the oviduct or fallopian tube. Hormones play a crucial role, so must be mentioned in your answer.

Page 61 — The Menstrual Cycle and Fertilisation

1 a) *Maximum of 2 marks available:*
No ***[1 mark]****, because the progesterone level has fallen, triggering menstruation* ***[1 mark]****.*

b) *Maximum of 2 marks available:*
Around day 12 ***[1 mark]****.*
It causes the release of the ovum (ovulation) ***[1 mark]****.*

c) *Menstruation* ***[1 mark]****.*

Page 63 — Pregnancy and Birth

1 *Maximum of 5 marks available:*
It secretes the hormones progesterone and oestrogen ***[1 mark]****.*
Oxygen and products of digestion diffuse across it from the mother's blood to the foetus's ***[1 mark]****.*
Waste products like carbon dioxide and urea diffuse from the foetus's blood to the mother's ***[1 mark]****.*
It stops disease-causing bacteria getting to the foetus ***[1 mark]****.*
Some antibodies can diffuse across, giving the foetus temporary immunity ***[1 mark]****.*

2 a) *Maximum of 3 marks available:*
Oxytocin stimulates the smooth muscle in the uterus wall to contract ***[1 mark]****.*
As more and more is released, the frequency and force of contractions increases ***[1 mark]****.*
Oxytocin also causes smooth muscle contractions which expel milk from the breast into the baby's mouth ***[1 mark]****.*

b) *Maximum of 2 marks available:*
HCG prevents the corpus luteum breaking down ***[1 mark]****, so that progesterone is still made and the endometrium stays thick* ***[1 mark]****.*

c) *Prolactin stimulates the breasts to produce milk* ***[1 mark]****.*

Section 9 — Ecosystems and the Environment

Page 65 — Modes of Nutrition

1 *Maximum of 5 marks available:*
Named parasite: Taenia solium ***[1 mark]****.*
4 marks for describing adaptations, from any of the following:
Hooks and suckers as a means of attaching to the host ***[1 mark]****.*
Covered in a cuticle, to resist the defence mechanisms of the host ***[1 mark]****. Many eggs are produced to increase chance of spreading to another host* ***[1 mark]****.*
No digestive system— it just absorbs the host's digested products ***[1 mark]****. Long and thin for a large surface area to absorb nutrients through* ***[1 mark]****. Long and thin shape so that it doesn't block the gut, killing the host* ***[1 mark]****.*

2 *Maximum of 4 marks available:*
Rhizopus fungi have thread-like cells called hyphae to penetrate the food ***[1 mark]****. Enzymes are released from the hyphae to digest the food surrounding them* ***[1 mark]****. Large insoluble molecules are broken down by the enzymes, giving smaller soluble products* ***[1 mark]****. These are then absorbed into the hyphae cells* ***[1 mark]****.*

Page 67 — Energy Transfer Through an Ecosystem

1 *Maximum of 2 marks available:*
The biomass is measured at a particular point in time, so only the standing crop is measured ***[1 mark]****.*
Sometimes there can be a lower standing crop of producers compared to primary consumers if they don't live as long, but have a higher reproductive rate ***[1 mark]****.*

2 *Maximum of 3 marks available:*
100 units passes into the producers ***[1 mark]****.*
10 units pass into the primary consumers ***[1 mark]****.*
1 unit therefore passes into the secondary consumers ***[1 mark]****.*
Don't let daft little details like these 'arbitrary units' put you off. It just means that they're any old units, it doesn't matter what, and this actually makes the question easier for you.

Page 69 — Recycling Nutrients

1 *Maximum of 10 marks available, from any of the following:*
If a question asks you to 'give an account,' that means you have to describe what happens in words. But it never hurts to include a diagram, which will make things clearer for the examiner and help jog your memory.
Mention the 4 processes that remove nitrogen from the atmosphere:
Nitrogen fixation by free living bacteria (Azotobacter) ***[1 mark]****.*
and those in root nodules (Rhizobium) ***[1 mark]****.*
Lightning ***[1 mark]****. Haber process* ***[1 mark]****.*
Mention the 4 ways nitrogen enters and leaves the food chain:
Uptake of ammonium and nitrate ions by plants ***[1 mark]****.*
Consumers get nitrogen by feeding on other organisms ***[1 mark]****.*
Decomposers release ammonium ions ***[1 mark]*** *when they feed on dead organisms* ***[1 mark]****.*
2 marks for explaining nitrification:
Nitrosomonas convert ammonium ions into nitrite ***[1 mark]*** *and Nitrobacter convert nitrite into nitrate* ***[1 mark]****.*
2 marks for explaining denitrification: Conversion of nitrates to nitrogen gas ***[1 mark]*** *by Pseudomonas and Thiobacillus* ***[1 mark]****.*
Make sure you know the difference between nitrification and nitrogen fixation, which often get confused. Nitrogen fixation is bacteria turning nitrogen gas into ammonium ions or amino acids. Nitrification is bacteria turning ammonium ions into nitrate.

2 *Maximum of 2 marks available, from any of the following:*
Combustion ***[1 mark]****.*
Respiration ***[1 mark]****.*
Carbon dioxide released by decay of dead organisms ***[1 mark]****.*

Page 71 — Human Influences on the Environment

1 *Maximum of 8 marks available.*
Fertilisers increase the nitrate and phosphate levels in rivers and lakes as they are leached into water systems from the soil ***[1 mark]****.*
This leads to eutrophication ***[1 mark]*** *where the following happens — Algae and other photosynthetic organisms grow faster due to the increase in nitrates and phosphates* ***[1 mark]****.*
These organisms die faster than they can be eaten ***[1 mark]****.*
This increases the biological oxygen demand (BOD) of the water ***[1 mark]*** *due to the increase in decomposers, which feed on the decaying matter* ***[1 mark]****.*
Their increased respiration depletes the oxygen in the water ***[1 mark]****.*
The lack of oxygen leads to the death of fish and other organisms in the water ***[1 mark]****.*
Examiners really like questions about the environment, so make sure you learn these pages extra carefully.

Index

Index